Winter

Notes and Numina
from the Maine Woods

Dana Wilde

North Country Press
Unity, Maine

Winter

ISBN 978-1-943424-67-2

LCCN 2021937907

North Country Press
Unity, Maine

This book is for Bonnie, Jack, and Silas.

Acknowledgments

Special thanks to Patricia Ranzoni, Tom Sexton, and Kit Hathaway for their ongoing, invaluable encouragement and support, literary and otherwise.

"February, Maine" is reprinted by permission from *Coring the Moon* by Kenneth Frost (Main Street Rag Publishing Co., 2014). The last stanza of "Trick Staircases" is reprinted by permission from *Bashō In Acadia* by Mark Rutter (Flarestack Poets, 2014).

Many of the essays in this book first appeared in the Backyard Naturalist column of the Central Maine newspapers and the Amateur Naturalist column of the *Bangor Daily News*. "Winter Moons" first appeared in *North American Review*. "The Great Bear in Maine" first appeared as an excerpt in *Puckerbrush Review* and then complete in *The Antigonish Review*. "Six Lists in November" first appeared in *The Big Picture* and was reprinted in *The Café Review* and the *Bangor Daily News*. "The Auroras of Spring" first appeared in different form in the *Barre-Montpelier Times Argus*. "A Spider in Winter" appeared in slightly different form in *The Maine Entomologist*. "Pluto on the Borderlands" first appeared in *Pluto: New Horizons for a Lost Horizon* (North Atlantic Books, 2015). "My Old Friend Leo" is excerpted from *Nebulae: A Backyard Cosmography*. Cover photo in Troy, Maine, January 2002, by Dana Wilde.

Contents

Preface

Last year a snowstorm unusually huge for mid-December buried the Northeast. Maine missed the main event. Binghamton, New York, where I spent parts of four snowy winters in the 1980s and then the '90s, got forty-one inches, its all-time record. Towns in western Maine got two feet. Portland got a foot and a half. In Bangor and at our house in Troy, just three or four inches, although it snowed hard enough to seem like more.

The TV meteorologists were giddy because they love bad weather that causes no more than inconveniences. The evening of the storm, reporters and their camera crews were snooping around outside to see what was or wasn't happening. On one residential street in Portland, they came upon a guy jogging in the snowstorm.

The reporter and the newsroom anchors were delighted. A man jogging along unplowed streets!

"That is *so Maine*!" the news anchor said, laughing. "Yes, that is *so* Maine!" the reporter on scene said.

Their delight seemed to involve a romantic view of Maine, held largely by people from, well, points south, that *real Mainers* are so tough and hardy in winter they *go jogging in blizzards*!

Soon I received a text from my brother Alan, who lives in Gray and kills bugs for a living and was waiting in his kitchen for the whole mess to just end already.

"Did you see the TV news just now," he said.

I knew where he was going with this. "The guy jogging in the snow?" I replied.

"Yeah."

"'So Maine'," I typed.

"That's not Maine," he returned. "Maine is a guy looking out his window saying 'That guy's an asshole.'"

Maine has changed a lot in the past fifty years. Which you are going to find out in this book, depending on which pages you end up reading. You can start anywhere, and end anywhere, it's up to you, because what you have here is a collection of individual efforts to cope with winter. And one of the recurring points of bewilderment is that winter in Maine is different in 2021 than it was in 1970, to use a round number.

We come to this generality on the evidence of personal experience and of scientific data. For example, in 1970, I'm pretty sure it would have been nigh-on impossible for a television news crew to find somebody jogging in a snowstorm. Probably snow-jogging did happen, because all the elements are the same — there are still snowstorms, ice, wind, cold, and people from away who don't understand these things very well. But it ain't something "Mainers" much did. Too cold and wet to fool around like that outside to no practical purpose.

But also, all these elements are various kinds of different from what they used to be. The snowstorms can be bigger, but snow itself scarcer. The ice is pretty ubiquitously thinner and shorter-lived. Spring springs later than it used to. (Yes, it's true.) The cold is still cold — but as old-timer scientist Bob Nelson over in Clinton observed to me around the time of that big storm: "In the ancient days of my Maine experience (early 1980s), I remember stretches of three to four days when the nighttime lows were below minus 20 F, usually in late January or early February. I was quite sensitive to this, since it meant the fuel in my diesel VW truck turned to jello in the tank, and I got to walk to work. ... *Haven't seen those horrific cold stretches, that I can recall, in the past twenty years.* But, it's only been in that time that, again in my recollection, we've had nighttime lows below zero in March. Going to be some kind of interesting to see what comes in the next twenty years ..." (My italics.)

Whether we like it or not, climate change is upon us. And Maine winters are a sort of climatological barometer.

So changes in the severity of the weather make up a recurring theme in these little essays. And you may notice other recurrences that seem downright repetitious—the pre-winter stillness, the disappearing and reappearing sun, the wheels of space, the once and future climate. I have tried to acknowledge this repetition in a few of the titles. But also, each entry tries to give different, hopefully clearer shape to the same frosty images that arise every winter. If you know anything about the sculptor Alberto Giacometti, you know that he made dozens of noses—the same basic form over and over again, trying to capture a nose. If I remember my art history right, he never thought he succeeded.

So it is with my winters, except in words, which is almost all you've got in January and February. Because the other thing is, winter's main feature is that it is almost featureless. It's vast and blank, compared to summer and fall when plants, birds, and all kinds of other creatures and processes are an abundant source of astonishment, natural detail, and beauty. In winter, snow and ice are everything. And darkness, which gives your imagination more than enough space to project words on the sky.

Of course, language is not all you've got, right? You can go snowshoeing, cross-country skiing, downhill skiing, skating, you can text, watch TV, snowmobile, ice-fish, trap, there's usually some shoveling you can do. Choose your own list of winter activities you think were of traditional interest to *Mainers*, before, let's say, about 1970. Around that time, the whole notion of who and what is a *Mainer* was beginning, like the climate, to change. I'm sure the guy jogging in the snowstorm was from Maine, like we say. But not the Maine my brother is from, who was ten in 1970 South Portland, or the Maine our enigmatic Grandfather Baker was from, who founded Portland Glass Company in the 1920s. He was a gruff old coot four generations along already from our ancestors in Boothbay when I knew him, and I think the chances of him ever "jogging" anywhere, let alone through snow, were exactly nil.

Winter. Maybe some of these thoughts will help get you through it.

After

Six Lists in November

the wind riffling dead grass
 waves in the blue water
 brown leaves hanging from bushes
 milkweed feathers:
the last exhalation of life, this time.

the wind bruises my nose
 the water beats a sailboat toward the beach
 oak leaves run like little kids in droves
 bright red berries falling off bare branches:
it turns & takes its last look at the cottage.

we used to hunt in these woods.
 we stepped on brittle twigs,
 we brushed through pine needles,
 we saw squirrels with their mouths full of acorns.
we teetered on the edge of winter with the bare poplars.

the cold clear air shows up the northern cross
 pindrifts in the sky, a bright one, Vega
 half the moon dumping light into the sea
 Orion's belt & another bright one, Sirius:
the late illumination of the sky this time.

thirteen geese in a chevron
 a gang of seagulls
 a chickadee among some naked branches
 four crows circling over a grove of pines
the last southern sunlight stuns their wings.

After

a black & red wool jacket heavy socks and gloves
 a sweater in the shade, a shirt for sunlight
 hats & freezing ears:
turn, turn, turn, change or fade, adjust or die; adjust.

November

Clarity

November's Gray Dissolution

Before snow falls in November, you can see right through the trees into the dead of winter and beyond it.

Early on in the month one fall, I set out on an overcast, wind-bitten afternoon to walk the railroad tracks along the west shore of Unity Pond. I wanted to see what the buttonbush brush looks like at this time of year. It was a gray, skeletal blur, like everything else. A little finer in the twigs than chokecherry, maybe. But on the edge of winter it comes to the same thing.

The bog between the tracks and the mouth of Sandy Stream was expectedly withered and brown. It was the exact natural reflection of William Faulkner's phrase "November's gray dissolution," which when I first read it forty-five years ago rang like a voice off Mount Sinai. It's echoed back every deer-hunting month since, as reliable as frost in October or the geese flapping and racing on before the snow.

Faulkner's story "The Bear" is about hunting, I was thinking while I calculated each irregular step from railroad tie to railroad tie, and part of the ancient tradition is getting hammered to ward off the chill, among other things. This reminded me that I had, when you got down to the bone of it, actually put myself in danger by walking exposed along these tracks in my gray autumn jacket. There was a clear shot at me with a deer rifle from the other side of the bog, and there's a sense in which it would be entirely my fault if somebody's optical equipment transformed the line of my slightly bent gait into horizontal brown fur.

A drug like alcohol can encourage this phenomenon, of course, but you don't have to be drunk for it to happen. Objective physical reality and your experience of it unify in your mind, and the biochemistry that makes up your brain has ways of reshaping reality that are to some extent under your control according to your inner strengths and weaknesses. It's possible for the mind of a completely well-meaning person with too much eagerness, too little inner discipline, and a rifle to turn a human figure into an ungulate figure. As, tragically, we know.

So foreseeing the possibility, however remote, of becoming a party to my own sudden passage into the next world, I turned around before I reached the trestle bridge, got in the car and ambled back out along Kanokalus Road, which runs right through the middle of a cemetery.

November's dissolution does not get any grayer than this. Its beauty is astounding. Lines of maples, birches, and ash who have gone bare for the long sleep. Apple trees with dark fruits but no leaves. Milkweed feathers and fraying cattails, leaf detritus, goldenrod ghosts, the shells of Queen Ann's lace, staghorn sumac naked except for small pyramids of dry drupes, dead grass, all of it reddish brown. Towering overhead, luminous yellow tamaracks. The sparrows of fall, the Canada geese in loose low-altitude chevrons. Even at midday the sun drops only in windless dappling on the ground. All that stays is dying, and all that lives is getting out, the song says.

It's a kind of magnificent desolation, to lift a phrase once applied to the surface of the moon, that telescopes your mind. The unstacked firewood from backyard to backyard is like a thousand other unfinished projects teetering on the edge of too late. Winter is not now, yet it will come.

November

Here and there along the roadside, through gray, bare, bony tangles, winterberries remain like red candle flames. *Ilex verticillata*, the arborists call the bush, black alder to the rest of us. Overwintering birds, raccoons, and mice will get the seeds when the better foraging has itself gone south. The fruits turn human stomachs, but in the mind's eye the startling red berries run tunnels through the gray twists and turns of deep brush, and transform November, there in the far invisible post-frozen distance, into the rosy underpresence of spring. The readiness is all.

Tamaracks

The natural sign of the last transition from fall to edge of winter, beautiful and melancholic, is the tamaracks. In mid-November they turn gold.

They're the only conifers in our part of the world to shed their needles in fall. Others—hemlocks, spruce, fir, cedar—drop a few needles at a time throughout the year. The old white pine blankets the driveway with fragrant brown needles in June. It can keep photosynthesizing whenever it's warm enough, instead of shutting down to wait for summer. But the tamaracks are anomalously deciduous, not evergreen, and close down for winter like the alders and birches, their neighbors in wet soil.

Most of the Eastern larches, as they're also known, in the lower forty-eight states grow in Maine, in slivers of New Hampshire, and around the Great Lakes. They reach fifty to sixty feet high or more, and live a hundred and eighty years or so, depending on site conditions. Western larches (*Larix occidentalis*) can top out at a hundred and fifty feet and live three hundred years; some that may be more than nine hundred years old have been reported in Washington state.

Though not generally as abundant as spruce, beech, fir, yellow birch, and some others, tamaracks were widespread in the Dawnland area of North America before Europeans arrived. In fact the word tamarack most likely derives from the Western Abenaki word hackmatack, which seems to come from the word *akemantak*, meaning "a kind of supple wood used for making snowshoes." The Penobscot word is *mɘ̀nɘhokak*. Up here in Euro-American times we call it Eastern larch, black larch, American larch, or *Larix laricina* to the botanists. In northern Maine it's sometimes nicknamed juniper, which keeps the scientists up at night because juniper is technically (and practically) a different plant.

The wood is tough and fairly good to burn, according to the U.S. Forest Service. The Indians used it for arrow shafts, and hollowed out large burls for pots. In Alaska it's used to make dogsled runners. Wooden boat builders used it to join ribs to deck timbers, as well as for planking. Dawnland Indians have used decoctions made principally from the bark to treat colds and coughs.

More to the point for a backyard naturalist, the tamaracks generate one of the final arboreal signs of the time. While the world closes down in November, beauty knells up through their branches on the edge of bogs and winter. It's almost a religious paradox. Those gold-needled steeples steering skyward, like a perennial Dome of the Rock in the northeast forest.

The seasons are the foundation stone of reality here, and the woods transduce its light.

Moons

A round Thanksgiving, November starts to leak into December and look like winter. There can be a crust of snow already edging the backyard inside the woods. It happens less in recent years. But November has always, and especially lately, had more of a distinct late-autumn personality, different from any other month, really.

Not sure why, exactly. They all bleed into each other to one extent or another. But in November, something about newly bare trees distinguishes the whole sense of the world from before Halloween, which is one thing, from December, which is another. The trees in this deer-hunting span look like a tangle of gray skeletons. Some marcescent leaves stand out, mostly oak brown and beech dull, and the tamaracks have gone to golden flares. They stand there, to quote an old rock and roll song, and seem for all the world to be just patiently waiting for what's inevitably to follow. The *Farmers' Almanac* calls a full moon in November "beaver moon." Abenakis call it "freezing river maker moon," according to one academic source, and the Passamaquoddys specifically "freezing moon."

Often, now, a freezing wetness feels like December descending early. It's usually after Thanksgiving, sometimes well after, when nights start to go cold for good, by which I mean consistently below 30 degrees. This is because of lowering sunlight, lower and lower for shorter periods every day. The *Farmers' Almanac* calls a full moon in December "cold moon" or "long nights moon," and the Abenakis "winter maker moon," which sounds exactly right to me. The

Passamaquoddys have "frost fish moon," which is outside my reckoning, really, never having been an ice fisherman.

If November is bare and December is the beginning of the end, January and February are the deep freeze itself, with distinctions maybe less obvious than November's skeletons from October's colors and December's hardening. In January the junipers are shagged with ice, which in normal times never really lets go, and the spruces look rough in the distant glitter of the low winter sun. January is the mind of winter, maybe summed up in names like "wolf moon" or the Passamaquoddys' "whirling wind moon," if you see what I mean.

In February the cold bites that much deeper and wider, so you get full moon names like "starvation moon" and "snow moon." Abenakis also have "makes branches fall in pieces moon," and Passamaquoddys, similarly, moon "when the spruce tips fall." But something else intensifies February that is almost impossible to notice in the ice of January, and that's that the sun, which has been grinding only very slowly higher in the sky, begins a steep upward swing in mid-February. March can still be cold at night but starts to thaw in the elevating sunlight, so lunar apparitions get names like "worm moon," because life is starting to emerge from underground, and Passamaquoddy simply "spring moon," and Abenaki "spring season maker moon."

But despite the tidy, one-to-one, month-to-moon-name charts you can find online and in books, reality is something quite different. The months have certain recurrent characteristics, but in truth they often do not fit neatly into preconceived categories. Month-naming is a human-made illusion. The moon haunts the very word: two thousand years ago in Northern Europe, "month" referred to a complete moon cycle from new to full, which lasts a sliver more than twenty-nine days. Time-keepers narrowed the named months down to twelve a year, but since all but one of our months has thirty or thirty-one days, it turns out there are usually thirteen full moons—thirteen months—in a calendar year.

So when the chart tells you that "harvest moon" or "corn moon" means September's full moon, or that "hunter's moon" means October's, or that "beaver moon" is November's, the chart is cheating. For Abenaki and Passamaquoddy people, phrases like "freezing moon" are not, from ancestry immemorial, talking about a period of time called "November" at all. They're talking about full moons that occur during a condition of the seasonal round.

In what might be more accurate expressions, a Penobscot website calls the moon corresponding to a full moon in maybe late October or November, simply "moon of autumn" (*takʷakəwí-kisohs*). A full moon appearing in the thirty-one-day span we cling to calling December is "moon when ice forms on the margins of lakes" (*asəpáskʷačess-kisohs*).

I like this a lot better. It looks just like the beaver pond up the road around this time of year. It relieves the burden of wondering why December is horning in on November, which is not a "month" at all, but a state of mind. A time of a moon of autumn, a freezing moon, a waiting-for-winter moon. I love the skeletal trees, the somber gray and brown, the golden tamaracks, the way they recur and settle in. The stillness here is all.

November's Bare Clarity

When the snow holds off in November, as it's done pretty steadily for years now, a late-autumn clarity crystallizes on the woods and fields. It's like looking through a lens. Every solid object in the landscape turns into exactly what it is.

The blue jays, I mean, are blue all spring and summer, of course. But from late October on, their blue, black, and white feathers gain a brilliance that didn't seem to be there before. Or somehow did not strike with such intensity. As if they absorbed the gathering darkness and turned azure into a silhouette.

The same thing happens in the trees. Perched crows stare east in black Giotto profile. All that's left on maple, birch, and ash, chokecherry and winterberry, are tangles of gray stems and twigs, sharper to the eye than the drips of an abstract painting, and more coherent. Oak and beech leaves hang on like copper medallions.

I think it's the low angles of sunlight that transform the remains of summer into November. Red and yellow October has devolved to amber, brown, and burnt sienna. The tamaracks turn a planetary yellow and seem to glow of their own accord, botanoluminescent even on a gray day. Firs and spruces throw long, deep shadows onto hayfields. A fading mown expanse backdrops reed grasses and goldenrods that expired in summer but now look like illuminated spirits. In the clear low sun, everything is waiting on winter.

Late mid-November afternoons the sun throws rays of light from less than 10 degrees of altitude over the horizon. At the same time on a mid-June day, its altitude is nearly 44 degrees. So the late-autumn rays come slanting through the atmosphere much shallower, but much deeper on the eye, for some reason. I think because at low angles the light's polarization—which humans experience as glare—is vertical, so the scatter of photons is less. I don't know. The light seems purer. And so do November's artifacts—trees, abandoned hayrolls, exploded cattails, red winterberries, slow-collapsing stone walls, birds, and stalks of dead grass.

Vertical is the angle of descent.

When I was younger than my son is now, I thought the greatest clarity obtained in August, when the humidity filters out and Canadian air makes its first presageful sweeps. Some years farther on it seemed like September gave the sharpest picture of reality, until, awhile into that hypothesis on how the seasons drive your mind, it started to seem like apple-picking time really is a gauze—of mists, and mellow fruitfulness. September sunlight is not rosy-hued, exactly, but gold in the afternoon. Now in recent years the chill is off October, too. The slant-lit milkweed and wild carrot skeletons by Halloween are signposts to the real turn: November. The monthlong moment when sunlight makes its final descent toward winter.

Has global warming put deep autumn off a month or two? Or are the shapes of things just coming clear that never were in youth? Does age congrue to time of year? You look back, and what you see—the frost perched on the ground, the blue jays stealing from the cat's dish on the step, the bare birches, the firewood stacked, the deer in rut, the gunshots, the pain in your hip—all looks exactly like what it is and always was, though you never knew it in the glare of the way up.

Winter's closing in, she said.

Still Point

One morning I was looking out the back door into the branch-bare, copper-colored, late-autumn quiet. This is exactly what November looks like, I was thinking. "The still point of the turning world."

Just as the words came into my head, light glinted from a straggling, yellow beech leaf directly into my eye. It was as bright as a star. It shone diamondlike for a long few seconds. Then flickered and faded. The backyard returned to still, brown, Saturnian November.

The mechanism was simple enough. A beam of light traveled from the sun at approximately one hundred eighty-six thousand miles per second to Earth. Earth rotated the beech leaf at about one thousand miles per hour into a clear path through tree twigs and branches. The light ray traveled a silk-thread-wide southeasterly path through the tangle of pine, spruce, fir, and deciduous skeletons and struck the dew-wet leaf. It bounced off the leaf at exactly the right angle onto the back of my eye.

This cosmic alignment lasted one still moment. The Earth continued to rotate. The stems and branches in Troy, Waldo County, Maine, North America, Northern Hemisphere, were rolled in Earth's diurnal course into the path of the sun ray. The star-leaf glinting on the back of my eye winked out. November's shadows returned. At the point where I was standing, that is.

The condition that allowed the light to strike my eye was November Earth. Because it's tilted by 23.5 degrees on its axis with respect to the sun, the Earth's poles angle toward and away from the sun once each in the year-long orbit. Now, the Northern Hemisphere is approaching its maximum tilt away. Light rays angle in lower and lower every day. They hit full tilt around December 21 each year, the shortest day of lowest sun and longest midday shadows. November and December sun rays filter through my backyard trees. In June they stream in unobstructed from high above the entangled oak tops.

Angles of light make November what it is. The somber colors carried on electromagnetic waves. Bare, shadowy trees. A leaf glinting at the still point of the turning world.

It feels like the cosmic truth. What that truth is, I can't say. Not the glint, not the leaf, not the dew, not the rotating Earth, not even the angle, and certainly not the words. Just the cosmic moment, glinting.

To keep up with the careening world, keep moving. To keep up with the cosmos, keep still. Suddenly, from out of the shadows, come light rays beaming into your eye, neither from nor towards.

November is an astonishing revelation, year in and year out.

"Isn't it winter, by then?"

Often, in recent years, summer seems to stretch all the way into November. Meaning there have been spells when the daytime temperatures at my house have included 60s and up to 70 right through the first week of the month. I have kind of mixed feelings about this.

On the one hand, I had no objection to October 2017 turning out to be the warmest on record in Maine and New England. In fact, according to the National Oceanic and Atmospheric Administration, the temperatures in New England from May through October that year were "much above average." Warm weather is welcome, on a day to day basis.

But if you can remember back more than a couple of decades, you also know there were reasons why Labor Day weekend was viewed as the end of summer. Not only did school start, closing down family vacation season, but fall had always decisively sharpened the air by then. I remember a nimbus-rocking northwest gale stiffening my young knuckles and blowing my ball all over a golf course in Naples one long-ago Labor Day. It didn't seem unusual.

By mid-September, frosts settled on the tomato plants and leaves were always turning. I remember forty years ago gazing out a classroom window in Orono about the third week of September during the first college course I ever taught, and thinking the red, yellow, orange leaves in the woods up the hill looked like a bowl of Trix.

In those days a spell of warm days usually returned in late September or early October—and by that I mean temperatures popping back up to 70 degrees or so. That's what the expression "Indian Summer" referred to. By the 2010s, Indian Summer no longer existed, really. Those few days or a week now make up the whole of September. The red maple by the driveway that twenty years ago was always aflame before month's end, now waits until October to turn. In 2017 it was ten or so more days further on. We've seen very little frost on the cars before October in recent years, here in Troy anyway.

Awhile ago I gave a reading to a group of fellow backyard naturalists, including a rumination on the turn to winter, and afterward one of them asked me why I referred to November as "fall." "Isn't it winter, by then?" she asked.

All I could think of to say was, "No." Not with Novembers past in mind, anyway.

Winter is characterized by snow and ice. It used to be that snowfall measured in multiple inches could be expected in November, and then your internal clock would re-set to winter. I remember at least one school-closing blizzard the first week of November in the late '60s. About twenty years ago my brother's pickup got stuck in the driveway on Thanksgiving Day. There was a hefty accumulation of snow in 2014, and there are still cold spells. But the average snowfall for the month overall is less than two inches, according to one meteorologist's blog. With a few exceptions, not much frozen stuff has fallen out of the sky in November for years. November is the tail end of autumn, not the beginning of winter.

This doesn't mean snow doesn't fall in November. It can still happen in its natural turn, the same way the giant hurricanes that wrecked Houston, Key West, and Puerto Rico, the fire-kindling drought out West, the disappearing glaciers in national parks, our unusual blow-down windstorm in October, all could be events just in their natural turn.

But you know, they're probably not. Probably they're all effects of the Earth's climate changing much faster and sooner than the scientists even predicted.

Which brings me back to that summerlike October, and early November. I like it in the short run, but I fear what it means for the long run. I have a little grandson who may end up with few, or no November snowstorms in his memory banks. I hope it is not any worse than that for him. But I don't know.

Late November

Just before a recent snowfall, I stopped along the walk I take most mornings and saw four ducks paddling around on the beaver pond.

The pond was slate-gray and flat, in that hunter's stillness November balances on. The air is chilled, but not yet wintry. The trees are bare. Their branches are gray and skeletal, and ragged white clouds knot in dark blue sky. Hardly a breath of wind.

I crossed the road and climbed up the short, steep bank of dead grass. Over white-green moss and juniper, under cedars and tattered spruces, I made my way to the edge of the pond, which is held back by the outcropping of ledge where I stood and a makeshift barrier of deadwood and leaves which is leaking. A hundred feet across the pond is a beaver dome of piled and interlaced sticks. By the far shore were the ducks. They looked like black ducks.

I watched them glide around on the shale-like water. From time to time one stuck its head under and after a pause came up spluttering and spraying beads of cold water around. They meandered in the direction of the beaver dome.

Suddenly there was splashing and wingbeats, and they all were airborne, rising like seaplanes. The strange thing about this is how they all spring together—not one after the other, or three following one who panicked, but all at the same instant. They flapped almost in unison and climbed smoothly over the water and the beaver dome, then the frost-crusted hayfield beyond.

About the time they crossed the shoreline, a great blue heron arose like an apparition from the reeds the other side of the beaver house. It was shaped like an assembly of joints, with sharp head and long neck, immense pointed wings and slate-blue sticklike body. It had the angles of a pterodactyl and the beauties dinosaurs lacked, and it was, amazingly, totally silent.

Its wings stroked slowly and powerfully, moving without pressure, almost, over the wet leaves in the autumn chill. It seemed to float through the air, headed westerly behind the ducks.

Weeks earlier in October, as I was walking up the driveway near the house, a motion over the brook in the fir woods twicked the corner of my eye. I hesitated and turned, and saw a gray-blue winged shape rise from the brook, waft up through the trees, and vanish. It was noiseless as the woods. It had the size and shape of a heron, but how could a bird that large navigate through hemlocks and pines? I wondered if I hadn't seen a woodland ghost leak from a crack between two seconds.

The heron over the pond also ascended in complete November silence. The pond surface was undisturbed, and all I heard for some moments was water trickling out through the stick dam and across rocks into the gully by the road. Everything had paused, as if taking one last breath before winter, and then was quietly gone.

December

Dark

Winter Frontier

As it finally grew cold this year, the vestiges of summer dangled like bits of grass and twigs in autumn's last ragged spider webs. A lone yellow hawkweed, contracted against the cold, looked up out of the grass by the gravel walk. A little viney beast with tiny white blossoms and heart-shaped pods grew near it—shepherd's purse, it was. There were dull orange marcescent oak leaves. Stiff willow-herb. Winterberry branches heavy with red berries, like a galaxy spun from seed. A murder of crows in the topmost branches of empty maples. Biting wind. Up above, two undulating chevrons of Canada geese honking in the cloud-strewn distance and flying due south. Lake Winnecook was as flat and gray as slate.

That night in the kitchen window two brown-colored spiders were hunkered down at the center of their orb webs. These two might have lived so long because of the unusual warm during the fall, but I don't know. They could not survive the inevitable snow, I didn't think, at least not in those exposed webs.

It's hard to identify many spiders to species. The orbweavers who live outdoors, many of them, die in the autumn but rise again in spring when their eggs hatch and a new batch of spiderlings takes over the age-old work. The webs of these two billowed and bounced together in the November gusts. One was constructed in taut, carefully measured rectangles radiating from the center. The other looked miskempt, with trapezoids loosely lashed to rough triangles. Maybe this spider was older than the other, less disposed to neatness. They were hunched down in the center of the webs, waiting for bugs that would never come. The wind batted the spirals up and down like trampolines, and the spiders clung there waiting. The silk is very tough.

They probably did not have long to live. Soon they would succumb to the cold and the natural cycle of their species. But if they lasted through the night, they would dutifully build again. They were like two old Chinese poets banished at the end of their lives to the northern frontier and gazing northward into places so bleak it is almost unimaginable. Cold, rolling, rocky grassland in the dark, with nothing beyond but more dark and grassland and strews of boulders and somewhere mountains. No town, no family, no tomorrow. Only vast, empty winter, in the end.

Sooner or later the snow will come. It will cover the gravel and willow-herb and goldenrod skeletons where the argiopes perished. How long the two old poets in the window will survive, I don't know. I haven't seen any new webs in the last couple of days. The shepherd's purse can live through snow. The oaks are almost stripped. Cold is coming to stay for what will seem like a lifetime. Winter is vast in northern China, and in Troy.

December Sun

One day along about the first of December with Christmas lurking just around the corner, I went for a usual late-morning walk up the road. No snow yet. Pickup trucks traveling past at roughly two-thirds the speed of light. Bite in the air, but no wind. Maples, birches, oaks, and elms bare and gray. Scudding clouds. The fields of the horse farm dingy with December. Beautiful, actually. Except for the vehicles.

It was getting on toward noon, but fir and spruce shadows littered the pavement. Through the skeletal woods you could see the glow where the sun was nearly as high as it was going to get. Tomorrow, it would not get quite that high, and the next day not as high as that. And so on into the month. To December 21, to be precise.

I looked at it too long, and the mottle out of the woods started playing tricks on the mind's eye. Almost midday, this was, and shadows splayed over the blue-gray road in the angles of summer evenings. Was it day or was it dusk, there in the morning of December?

Why this should be disturbing, I didn't know. But it was kind of haunting and bewildering, the sense of initial lostness you daydream might come when you're finally on the threshold of expiring. Or across it. What must ancient people have felt when the sun for all the world was sinking, day by chilly winter day?

Because I, of course, had no-nonsense science to fall back on. I rehearsed it in my mind as I walked and trucks blew by, and soon the unmistakable fear and trembling that was lurking in the sunlight behind the trees dispersed. Just to rehearse it again: the sun over the course of three hundred sixtyfive days appears to rise through springtime higher and higher in the sky to its highest point in the latter of part of June, and then descend through summer and fall lower and lower in the sky to its lowest point in the latter part of December. The pattern repeats exactly the next year, and the next, on into seeming eternity.

This is quite a strange phenomenon, when you dwell on it. Shadowy. But mechanically explicable. The Earth orbits the sun once every 365¼ days. It's tilted on its axis, meaning the North Pole and South Pole are not straight up and down with respect to the sun, but tipped at an angle of about 23.5 degrees. In the part of the orbit when the North Pole tips toward the sun, the sun appears to climb higher in the sky and stays up longer—summer. When the North Pole tips away, the sun appears lower in the sky each day and sets sooner—winter. (The Southern Hemisphere's seasons are opposite ours.) So this tilt, combined with the Earth's orbital motion, changes the altitude of the sun—as we see it—over the course of the year. It reaches its highest point around June 21, and its lowest point around December 21.

If you lived two thousand years ago, you had no access to these mechanical facts, although some ancient astronomers did have an inkling that even though it looks for all the world like the sun circles the Earth, the opposite might actually be happening. But in the ancient everyday, you had this mysterious annual phenomenon of the sun climbing up the sky and then descending down, year in and year out. Whose lengthening shadows can seem strange in December morning chill.

Your everyday knowledge of natural phenomena would be scientifically null but personally alive. The world was by and large experienced as a living activity then, however strange it seemed. A much different experience from the inert, objective clockwork our minds have learned to eye in about the past five hundred years. When you noticed the sun lower in the sky, you had the direct experience of it in the feeling of strangeness, mysteriousness. Instead of numbing your fear with a mental diagram of orbital mechanics, your way of talking about it was to tell a story: the sun like everything else in the universe is a living being, retreating in fall to leave us in the cold, and returning in the spring to warm the fields and crops and people. Forever.

Scientists usually call this story an "explanation," as though it was a misguided scientific hypothesis about a mechanical process. But that's not what it was. It was an evocation of the feeling that connects you to the phenomenon: the gigantic awe, mystery, and strangeness the yearlong rise and fall of the sun inspires in sentient beings.

Sun god stories still affect people today, in their many mythological manifestations. Because even though we now know how sky mechanics work, that sense of strangeness and awe, if you pay attention to the shadows, still exists. We are still human.

I don't mean science is wrong, of course, nor uninspiring. I mean it is one side of a coin with at least four dimensions. The imaginations that visualized the motions and mathematics of the Earth and planets revolving around the sun were prodigious. If you don't think so, try it yourself. Start with the practically undeniable observation that the sun rises in the east, moves slowly across the sky, and then sets in the west every day. Clearly, to your own repeated witness, the sun circles around the Earth. How would I—who still sees an M.C. Escher drawing every time I open the hood of my car—ever figure out it's the Earth that's moving around the sun?

You can see why the emphasis in ancient times was on the feeling you got when summer evening shadows plastered the horse track on middays in winter. It was a feeling of prodigious invisible power and strangeness. It was direct knowledge that the sun would flawlessly return. It was a sign of your own place in eternity. Where you too would return.

Scientifically, it is pretty sure—almost as sure as the Earth circling the sun—that Jesus was born in March. But your connection to him is his promise of the god's, like the sun's, return, so really, the truth inside the story is that his real birthday is toward the end of December.

I throw this out there as a suggestion from the edge of a December morning.

"There's always ground cover"
(Part 1, 2006)

Every year we wonder if we'll have a white Christmas, and every year we do. At least, that's my recollection.

In early December the ground freezes, usually, or it always used to. The grass becomes a carpet of brown spikes, and sleet or slushy snow might fall and then melt off. When we were kids this tormented us—would Santa's sleigh have anything to run on? But by Christmas Eve, snow always fell and blanketed spruces and backyards. Every Christmas Day in the 1950s and '60s was white, as far as I remember.

I also remember being warned not to skate in mid-December but doing it anyway. By then every pond at least *looked* safe, and anyway, there would be cracking and bubbling to tell you if it wasn't. In the 1970s we always skated before Christmas. And that was in southern Maine.

In recent years, though, the pond up the road from my house in Troy has barely a skim of ice by mid-December.

Making sense out of the weather is at least as tricky in the long term as the TV meteorologists find it one day at a time. And the facts of climate change can really be head-scratchers. By virtually all scientific accounts global warming has kicked solidly into gear.

In my unofficial memory, the snow cover is decreasing. Some winters in the 1960s the snow was so deep in Cape Elizabeth that we climbed on the roof of the house and jumped into the drifts. In December 1973, a friend and I were anxious to go snowshoeing, but there was hardly a dust of snow yet. My friend said to an old backwoods guy in Buxton, "I wonder if there's going to be ground cover this year." The old guy scoffed: "There's always ground cover." Meaning that from December to March, you need boots. I'm not sure that's true in southern Maine any more.

In the 1960s my father kept records of the date the ice went out of Great Pond in Cape Elizabeth. It usually happened around the first of April. Now, Great Pond doesn't even freeze solid enough to skate on in any given winter, according to old friends there.

I'm still pretty confident Christmas Days will be white in Troy. But we won't be skating outside until January. I wonder what comes next.

"There's always ground cover" (Part 2, 2020)

The mental adjustment you have to make every year from summer mode to winter mode is always wearing. But it seems to be getting more complicated for reasons, it turns out, beyond the late sixties, so to speak. Before it snowed this balmy December, I noticed myself hoping it wouldn't snow this winter.

What?

The feelings that went with this counter-reality realization were complicated. First, disappointment, because I've lived in Maine most of my life and know, of course, that in winter, it snows. Second, dejection—three to four months were still to pass before we could reasonably hope snow would cease accumulating. Third, what the hell was I thinking?

The fact that I would even have such a thought is alarming. Surely it signals a certain detachment from reality, or at least a state of mental denial that would be downright dangerous in February, when the low-slung afternoon sun is slowly wringing the neck of the geranium in the window, as Wilton poet Ken Frost so aptly stated it. As an old-time backwoods Mainer told me with absolute certainty one bare mid-December day about fifty years ago in Buxton (well south of Troy), "Don't worry about white Christmas. There's always ground cover."

I own my craziness. But, maybe it's not just me.

For one thing, didn't winter always used to run from No-vember to March? It seems like I remember Novembers with brutal cold and huge snowstorms—not always, mind you, but often enough to be ready for it. In recent years, though, Octo-ber has pushed further into November, the meteorologists don't seem bothered by runs of 50-degree days, and it's en-tirely reasonable to think it might not snow until December. And on the other end, February crashes cavalierly on into March every year lately, and March shoves its nighttime ice and daytime slush and mud and occasional blizzard into April.

This is all my imagination, right? Well, it turns out, it's not.

First of all, there is a clear-cut statistical correlation be-tween November and the rest of the winter, which one local meteorologist calls the "November Rule." When November temperatures are milder, then winter temperatures tend to be milder and snowfall totals lower. These tendencies hold 60 to 70 percent of the time in Portland and Bangor, and at slightly lower percentages in Caribou.

Next, there is the documented climatological fact that the Northeast, indeed most of the Northern Hemisphere, is on av-erage significantly warmer now than it has been historically. Most of Maine has reached average annual temperatures that are 1.5 to upwards of 2 degrees Celsius higher than the aver-age temperature since 1835. A USA National Phenology Net-work report of March 2013 states that: "Mean temperatures in [the Northeast] are expected to increase 1.4—2.2°C (2.5 to 4°F) in winter and 0.8—1.9°C (1.5 to 3.5°F) in summer over the next several decades."

Next, spring really is arriving later. Maps developed from the USA-NPN's analyses of data on leaf and blossoming times of certain plants, weather, and seasonal movements of animals show that in most of Maine, the spring of 2020 arrived much later than it has historically. It was the same story over much of the Northeast and parts of the Great Lakes region. (Farther south, spring arrived earlier than usual, much earlier from the Carolinas down to northern Florida.) The phenology—a component of ecological science studying seasonal and cyclic natural phenomena—indicates that from 1981-2010, spring bloom has been occurring up to three weeks later than it has historically in parts of the Northeast and Midwest.

This scientific description accords with my own unofficial grouchiness about April in recent years. Before about 2010, our dandelions showed up like clockwork on May 1. Lately they haven't popped before mid-May. It's hard to find clear year-by-year data for temperatures, but I'll bet anything that late-blooming springs occur when March invades April. And when October invades November.

What's happening, exactly, scientists are reluctant to specify. Except to make the very general observation that this kind of disruption looks a lot like what you'd see during significant changes in climate. And not to put too fine a point on it, but the Earth's climate is well into the process of changing significantly. And this is due directly, of course, to the forty billion tons of carbon dioxide we puke into the atmosphere every year. (The pandemic year of 2020 showed the world puking much less carbon dioxide because of less transportation activity. But it's not going to slow by one inch the climate change that is speeding downhill at us like an enormous mudslide.)

So by common wisdom, my hope that it might not snow this winter was ridiculous. There's always ground cover. But not totally unfounded. If the atmosphere continues to heat up like this, and adds roughly 3 more degrees Celsius in the next seventy-five years—which climate models show is not implausible—then New York City will be on average as hot as Bahrain. It would not be snowing here, in that climate.

I concede. It really was demented to hope it doesn't snow.

Poirier's Wolf

When I used to go camping, it was often with my old friend Phil Poirier, who grew up in Rumford and now lives in Farmington with my sister, Heidi. He taught me all the basics of carrying water and not falling off cliffs, etc., and went on over the years to become an active and well-respected member of Maine's hiking and outdoors community. He's hiked and guided excursions in many parts of Maine, and explained winter camping on Bill Green's TV spot. He has, as well, enhanced my understanding of the wilderness and wildlife.

Phil has a lot of thoughtful camping stories, of course, and one about an expedition to Canada seemed particularly worthy of retelling. So here it is, in his words.

"Years ago when I was younger and went winter camping a lot, I used to gather up friends to mount trips way north, to find the old-fashioned Maine winters we were already starting to miss. On a map of Quebec we noticed the huge doughnut-shaped Manicouagan Lake, which it turns out is a meteor impact crater forty miles across, three hundred miles or so due north of Fort Kent. And close to it was a road that had JUST started being maintained for winter travel. OK!

"We rented a crappy minivan to fit all our stuff and set off on the eighteen-hour drive in late March, hoping to catch the milder tail end of winter (meaning warmer than minus 30 Fahrenheit at night). We brought topo maps of the area, but knew little else. Map, compass and streams would guide us into Les Monts Groulx.

"I had picked a camp site a day and a half's hike in, on the outlet of a pond that would be likely to have open water. Turns out I nailed it! The site had a great open southern exposure, with a view to the open summits of the low mountains we were to explore. We had glorious day hikes around the area's lakes and mountains. We were in taiga, which means that the trees are pretty stunted, mostly black spruce and sparse, so it was easy to navigate with map and compass. A tree fifty years old might only be three inches in diameter. We were in the woods for ten nights, and of those nights we saw the Northern Lights seven times. Each day on the trip was mild: 20s during the day, around 20 below at night.

"Late in the trip, I decided to walk out onto the pond one night to watch for the aurora. Walking away from a fire or my companions to stare into the night sky is one of my true joys, and this was no exception. The Milky Way spread like a gauzy veil, and I stood staring into that expanse for a good half hour. Normally, standing in below-zero weather will get you pretty cold, but not this night. This night I felt the power of the Earth surging up through my feet.

"I don't know if you believe in chi, or prana, but I was in a state of such deep meditation, I believe I had summoned my chi to stay warm, motionless, staring up into the guts of space. After a half-hour I thought I should move my neck or it would become sore. I rotated my head, and immediately, right nearby on the snow-covered ice with me, came a deep woof. Hair stood up on my neck (and did again as I wrote this). I couldn't see anything … it was too dark. Immediately I started walking back to camp, trying my best not to run and betray my fear. Back at camp, my friends were skeptical of my story, thinking I was trolling them.

"But the next morning, we found tracks all around the perimeter of camp. LARGE tracks. This was no coyote.

"So of course I had to go back to the pond to see how close the animal had come to me in the night. Sure enough, the tracks showed a wolf had been walking in our trail, which bisected the pond. It had come within thirty feet of me. Then it made a right angle beeline for the nearest point of land.

"I think the wolf had just been tooling along the pond while I, meanwhile, was downwind, standing stock still. The wolf was either curious to see what I was, or maybe just on autopilot, then saw me when I rolled my neck.

"A wolf in the Canadian wild. It's one of my most treasured experiences in the Great White North!"

The Winter Sky

Just around the time the holidays close down, that apparently interminable stretch from December to March is peering over the northern horizon with dragon eyes. For months it will be too much trouble to clamber over walls of plowed up snow to get to the Shed and its creaking floorboards, only to forget what brought me there in the first place.

In a way the iciness of deep winter lives in the idea of "north." It's where the cold comes from, and certain stars reflect it year round, especially Polaris, which marks almost exactly true north. Even in summer it looks bluish-white and icy up there. Who knows which came first, the cold white star or the cold white apprehension of inevitable winter.

Anyway it doesn't matter because a lot of northern stars will do for the idea of cold. Ursa Major—the Big Dipper—wheels around the pole all year, never setting, in tandem with Ursa Minor, the Little Dipper, where Polaris is the tip of the handle. They shine brightest on cold nights, and have done so for longer than millennia. Polaris is the central axis around which the whole sky turns.

The central axis for now, that is. Similar to the way the seasons eternally turn, the stars also turn, not just nightly from east to west as the Earth rotates, and yearly, as the Earth revolves around the sun, but also in a much longer cycle where the northern (and southern) stars travel a great circle. Polaris, in other words, was not always the pole star. About twenty-five hundred years ago, Kochab, one of the bright stars in the Little Dipper's bowl, was nearer true north than Polaris.

"True north," in my parlance, is at the top of an imaginary line or axis running through the Earth perpendicular to the sun. The Earth's spin, however, is not exactly along that axis. It's tilted about 23.5 degrees. In other words, the North Pole does not point at true north—it's 23.5 degrees off the perpendicular to the sun. This tilt gives rise to the seasons, as our hemisphere tips into the sun's warming rays in summer, and away from them in winter.

And not only is the Earth tilted, but it's slowly wobbling, like a gyroscope knocked off its up-and-down balance. The wobble is called precession. One full wobble takes a bit less than twenty-six thousand years. This means the northern stars appear, over these thousands of years, to make a great circle. So every few thousand years a different star marks north.

In our time, Polaris is within a degree or so of north. But around 500 B.C. (twenty-five hundred years ago), Kochab was nearer north. Around 3000 B.C. (five thousand years ago) Thuban, in the constellation Draco, the Dragon, was the pole star. About a thousand years from now, the star Alrai in the constellation Cepheus will mark true north. In 14000 A.D., Vega will be within about 5 degrees of north. And after one full wobble, in 27800 A.D., Polaris will again be the North Star. This has been going on for countless millions of years, the result of torque from the sun's and moon's gravitation on Earth's bulging equator.

How long ago humans first noticed the great slow star-circle, forgot it, then relearned it, no one knows. The entryway of Egypt's Great Pyramid, which seems to have been built around 2800 B.C., aims exactly north, which in those days would have been marked by Thuban. This is not evidence the

Egyptians knew about the great year, necessarily. But records or handed-down memories from that time indicating the positions of stars would, after a few hundred years, reveal the stars were changing position in slow circular motion. Plato in about 370 B.C. referred to the great year. And some scholars think mythological stories contain coded information about the stars, the great year, and more. Stories that could be fifty thousand years old or older, some of which are still remembered today but whose figurative codes are forgotten. It may have been foreseen that Thuban ages hence would return to north.

Thuban is a strange star, in the view from my driveway. It's the alpha star of Draco even though it's not the brightest in that constellation. Its name is Arabic for serpent, which makes sense, but it came by it through repeated medieval copying mistakes from the Arabic phrase *r'as al-tinnin*, the serpent's head. It's classified as a white giant, a type of star rarely seen. It's cold-looking. Distant-seeming even beyond its faint proximity to the bowl of the Little Dipper, like an echo of something, a cycle of time so slow it was forgotten. It's lean and hungry, and projects a boreal chill.

Or am I projecting my own chill? I don't know. I can't even remember how I got here. I only know it's winter again, the trees and branches will be cracking, and moonlight soon will refract through icicles and frost, while Thuban and Polaris, as ever, watch.

The June in Juniper

For a long time the scientific compartment of my mind has explained to me that festivities exist in and around December in order to help everybody buck up for the months to come. The sun is at its lowest high point in the sky and winter is about to set in hard, so we might as well celebrate the recent harvest and the remnants of our summer cheer to gain some momentum into the long, cold dark. This turns on the hope and faith that light and warmth do indeed return.

Hope and faith are—not to put too fine a point on it, as wires between the scientific compartment and the mystical compartments start to cross—psychological activities, and "psychological" is somewhere on the outskirts of spiritual. Whatever "spiritual" means, you nonetheless need a mind of winter to see the junipers shagged with ice and not let it halt your hope of spring. Especially in the darkest season of the year, you need to give attention to the evidence of the thing unseen, which in December is decidedly June.

For example, there is abundant empirical evidence, even in heavy snow, that life survives winter: the cedars, spruces, firs, and pines crusted with snow in January are still green. The deep green that doesn't fade and symbolizes life: in ancient times our forebears used evergreens to keep in mind that even in the deepest depths of cold and dark, light and life return. That's not misery you hear when the wind whips through December hemlocks, but the underlying whirl of life.

And they noticed that the shortest day in December is exactly when the sun begins its return. That's one thing evergreens like mistletoe and holly mean, and why they're used in wreaths around the holidays.

Holly, which has shiny, leathery, pointy evergreen leaves and distinct red berries, is not native to Maine and in truth I've never seen it growing here. Our mistletoe also is not the mistletoe of the British Isles associated with healing, good-will, and eventually a kiss (our Eastern dwarf mistletoe is a parasite to spruce). What we do have in abundance, though, is juniper.

Common juniper, even more than balsam fir, smells like Christmas to me. It's the most widespread evergreen in North America, and around here grows mostly as flat bushes in rock-ledge places. Its needles are stiff and pointed, and Native Americans apparently burned them as incense and a cure for coughs. Juniper berries—which actually are tightly closed cones, ripening from green to dusty blue over two or three years of growth—have a sharp, piney fragrance. A regenerative tea can be made from them, and centuries ago the Dutch discovered they create a pleasant, biting tingle when included in the distilling process for gin.

To me gin also tastes like Christmas. This is an association blowing around in some bare place outside my scientific mind, not a fact. And I never drink gin to get me through the night, like some people. But in scent and effect it might be the quintessential "glass of cheer." A spirit of the spirit of Christmas arising out of evergreens.

For the junipers and their berries, stark in the disturbingly frequent snowless December terrain, seem as gone as the way-worn sun. The scientific naturalist in you, though, knows that the tough, windy little bushes don't actually die in winter. And on some warm June afternoon when you walk by at just the right moment, the juniper scent will rise and suddenly it will be crystal clear to the backyard naturalist in you that the Christmas hope was true after all, you survived, and life prevailed, as promised.

A Place Where Houses Are All Churches and Have Spires

One December afternoon before he outgrew every shoe in the house, my son, Jack, and I went to the woods to cut a Christmas tree. He was old enough by then to navigate the snow the same way I did when I was a boy hunting blue spruces, which were pretty elusive in southern Maine.

We hiked up the hill behind the house through floating snowflakes with an ax and great expectations. We examined small hemlocks and pines, and ragged black spruces. Chickadees were living in the big ones. Some white spruces were the right height and had fuller needles but smelled like skunk. We saw a lot of sparsely equipped young firs. Our boots crunched snow for a long time.

In my boyhood, the woods seemed to have a reverent air, like a cathedral. At Christmastime the church pageant, colored lights, and the likelihood of Santa angling his sleigh through the treetops at night blazed in my mind. The balsams and spruces were a place where houses are all churches and have spires.

Commercially it's called "magic," though there might be a better word for the depths that underlie the advertising. Christmastime is older than Christ.

For Christ (the scholars tell us) was probably born in the spring, not December. But long before his time, rituals of rebirth and renewal were celebrated at the winter solstice, the day in late December when the noon sun stops being lower each day and begins getting higher. To ancient people, this was a powerful cosmic moment. The sky and its workings were expressions of activity in the divine world, and the sun lowering in the sky from June to December, and then rising from December to June, was a living, awesome sign of the power that drives the universe. They revered the sun's rebirth, and used evergreens as ritual symbols because their needles, like the sun, survive and live again.

The profundity of this is hard to grasp here in the scientific age. As far as we know, we've outgrown childish superstitions about supernatural powers and our wise men understand that physics and biochemistry explain virtually everything.

As it happens, the pagan reverence of the solstice never really disappeared. During the 300s A.D., the emperor Constantine and the church fathers linked Christ to the renewal celebrations by appointing December 25 as his birthday. Christ, after all, is the personification of rebirth. We still use evergreens to remind ourselves that something awesome is happening just the other side of the sun.

On our tree hunt, I was hoping Jack would catch the drift of this, but he was getting tired. So we cut the fullest-branched fir we could find and dragged it home in the snow. Things are almost always larger in imagination than in the living room. The tree turned out to be a scraggly, Charlie Brown-style skeleton. We hung our ornaments and lights on it anyway, much to the shock of a delivery man who caught a glimpse of it, dropped his jaw, and scrammed to his truck.

Still, following in the footsteps of his wandering dad, Jack tells me he remembers that expedition, and the Christmas woods remain more real in his mind than all the scientifically grown and nurtured trees we've bought since.

January

Night

A Morphology of Winter

Winter is icummen in,
Lhude sing Goddamm.
Raineth drop and staineth slop,
And how the wind doth ramm!

These lines by Ezra Pound have inhabited my little world for more than forty years. He actually wrote them more than half a century before I ran across them on a blustery January day. And it turned out he stole the whole tune from a seven- hundred-year-old song about summer—"Sumer is icummen in, Lhude sing cuccu" (Summer's arriving, the cuckoo's singing loud).

What a sweet, raucous little poem to read on a bitter winter afternoon, when the world is made only of snow and words. One year, on the advice of a correspondent, my winter reading included a book Lew Dietz wrote more than forty years ago, *Night Train at Wiscasset Station*, which is a rumination on what Maine used to be like before the pivot out of the 1970s, when it was clear to him, me, and everybody else who could think in a straight line, that Maine was changing, possibly not for the better.

After most of a lifetime of frittering and digging around in English and some of its more abstruse contexts, and simultaneously devoting hundreds of life-hours to the complicated problem of winter in Maine and getting almost nowhere, I came across a word in Dietz's book that I had never seen or heard before: "brumal." It means: of, relating to, or occurring in winter.

Where was the word brumal when I needed it decades ago? It's as if it was buried in a snowbank with the other stuff left in the yard and forgotten back in November or October — a baseball glove, a bike, a length of chain, a life jacket, some of which you don't even remember owning. Then the snow melts and exposes it, and you say: That was there all winter? Where did it even come from?

Where did the word brumal come from? Well, it turns out to be directly from Latin, *brumalis*, a word derived from *bruma* (winter), which likely came from *brevimas* (brief), as in *brevimas dies* (the shortest day, i.e., the first day of winter).

Our word winter is straight out of Old English, a Germanic language that didn't start absorbing Latin cognates until 1066 when the Old French-speaking Normans beat the Anglo-Saxons at the Battle of Hastings in southern England. That happened in October. Bruma never touched winter, which not only survived through Chaucer and whoever composed "Sumer is icummen in," up through Shakespeare and into twenty-first century New England, but also goes back in etymological time to a cognate in eighth or ninth century Old Icelandic, *vetr*. Seven or eight centuries before that, the Gothic language had the word *wintros*, which likely derived from an earlier proto-Germanic word that was probably something like *wentruz*. And the linguists think *wentruz* probably came from a version of an original Indo-European morpheme from five or six thousand years ago, *wed-*, which is the ancient precursor of our words wet and water.

"Raineth drop and staineth slop!"

In Dietz's book, photos by Kosti Ruohomaa show bleak, snowy back roads and buried fields, chilly-looking Maine farmhouses, and old geezers from the 1940s and '50s with pipes sticking out of their mouths sitting around stoves and telling stories. Lobster boats in winter gales and seiners in hip boots pushing around sloughs of fish on cold-looking decks. I recognize all this — it was real. But I think that, while fishing and storytelling are still going on, what's happening in those photos is mostly gone now.

Ruohomaa, a Finn whose ancestral winters are even darker than northern New England's, sometime in the 1940s froze the Wiscasset train station in a moment of time. On a frigid evening, an electric light peeks feebly out of a rail car stopped near a snow-crusted tidal lip under a graying winter sky. This is what brumal Maine actually looked and felt like. Bleak, bone-chilling, and so deeply familiar of my childhood it's almost like the voice of the beloved itself.

Winter goes glacier deep in Maine. In memory, snow buries everything. On some January afternoons, the only things breathing are words. A raucous poem about winter. A book bearing witness to a passing era. It's cold so far back in time that even the Passamaquoddys, who were here for eons before English, Finns, or even Vikings arrived, don't exactly name it, but use a morpheme, *pun*, in agglutinative words that express whole sentences instead of isolating one bleak, inadequate sound to represent the whole multi-complicated season—*pecihpun*: winter has arrived; *puninaqot*: it looks wintry; *puniwi*: brumal.

Dietz's chapter on the seasons opens: "Responding to a question from a Danish Arctic explorer a half century ago, the Eskimo shaman said: 'We fear the cold and the things we do not understand. '"

When Dietz was saying "a half century ago," the Eskimo, in turn, was no doubt echoing thoughts uttered by Inuit ancestors who probably trekked the ice of the Bering land bridge twelve thousand years before that.

It was strange in that January's bewildering rainstorm to read Lew Dietz's 1970s ruminations on Maine's morph from the cozy ferocity of winters in the twentieth and nineteenth centuries toward whatever it is we have now.

Dark and Reckless (Part I)

Winter is a dark and reckless thing.

Where I got this sentence, I don't know. It recurs at some point every year after snow has fallen. More than six decades of relentless Januarys and it's almost all I have to say about winter. What else is there? Winter is blank snow. It's a torpor. It's an insulator. It's the yin of summer's yang. It's spruces frozen in the distant glitter of the January sun. It's nature's living figure of emptiness. It's death.

It's bitter. I remember reading long ago of eighteenth century settlers fleeing Maine-ward to escape social and political oppressions in Massachusetts and many of them bouncing back because of "extremely harsh winters."

Extremely harsh winters.

As early as the 1970s we were watching the newspapers' daily reports of winter degree-day numbers rise, year over year, and growing antsy about global warming. Even then. In the winter of 1983, it hardly snowed at all in Portland. Life consisted mainly of clambering over spits of sidewalk ice. That February I overheard a young bespectacled man, newly emigrated from points south, pretend to shrug off his visiting friend's complaints about the cold. In his confident, wise-be-yond-his-years mid-Atlantic accent he explained, "We've had a really hard winter this year."

I raised my eyebrows. My native-Portland companion returned the same quizzical look. We knew something about "extremely harsh winters," and this had not been it.

January

My siblings and I used to jump off the roof of our ranch house in Cape Elizabeth into snow drifts. Snow that deep, and often. We skated on Great Pond in December. It was snow-covered in February. My father kept a log of ice-out dates; I remember them being usually around the beginning of April. In recent decades, I've been told, Great Pond often doesn't freeze at all.

No doubt, what you call "a really hard winter" is relative. In Shanghai my Chinese colleagues thought it was hard when sidewalk puddles froze on a January night. An "extremely harsh winter" is something else. The settlers on St. Croix Island in the winter of 1604 experienced full harsh. Half of them died. Full bitterness. The full darkness and recklessness.

You can still sense the outer edges of harsh here in the early 2020s, if you've looked out the window at enough shadows and enough blizzards for enough years. The yards and driveways that look cozy and trim with pine needles and tamaracks and still-green grass in October, turn Christmas-white for a day or two in December. Then the snowbanks start to decay, and bleak and hollow follow like a hammer blow. The sweet autumn havens disappear, and in their places rise kerosene-smelling, blear-eyed, lonely chunks of interplanetary grit disguised as human dwellings. Not every day, but some. Enough. Even the inside of your own house can feel like it's inside the shadow of a crater on the moon.

Winter sun
with its
glass gloves
wrings the geranium's
neck
slowly
slowly.

So wrote Kenneth Frost of "February, Maine." He spent his later years in brumal Wilton. Worst of all is the hangover of waking-up March.

All you can do is watch shadows crawl across the driveway, the blizzards obliterate it, the blizzards of words that keep meaning almost nothing, it's so cold. Watch and wait. Sometimes the winterberries last a long time. Beautiful red droves of them in roadside bushes. They're among the last of the fall forage to be cleaned out by overwintering birds. I guess winterberries probably taste bitter to a bird. Lingering, maybe.

We haven't had the bitterest cold, 20 to 30 below zero, in Troy in this millennium. That I was awake for, anyway. The day-melt, night-freeze cycle we get instead, requiring bucket after bucket of salted sand on the driveway, is harsh enough, at this point.

It goes on for so long. In January, most of it is still ahead. The paucity of light, like shortness of breath. The silent ice. The back-biting wind. Dirt-crusted snowbanks. The bald eagles roaming empty space over Carlton Bog. You have to have a mind of winter to live in this kind of cold and not think of misery in the sound of the wind, so said one of the most disturbing poets in the American language. I'm still not there yet. It isn't misery I don't see. It's bitterness. The bitter cold. The bitter length of it. The darkness. And recklessness.

The great question is what will follow.

Aliens

In January with the holidays behind us, our minds turn to the grueling task of making the journey through deep winter. It is a labor like no other. It includes not only shoveling snow, some dicey drives home, and the thermometer registering so low in the minus-whatever range that you won't want to look at it until the sun hits it tomorrow afternoon. It will be awhile before the length of daylight seems like longer than a lunch break.

Your mind has to cope with it somehow. I am not making this up.

Depending on your ideas about what the cosmos is made of, winter brings not only eternally long night, but with it, moisture-free air that materializes crystal-clear skies that your mind can use, if you want to go there. Balls of thermonuclear night-brightness fire beam light down into your transparent eyeballs and trigger feelings of enormity, minuteness, awe, kaleidoscopy, and possibly divinity, grace, or terror.

You don't really get to pick, when you're standing there under Orion's southern-sky-spanning belt, scabbard, and bow on winter nights. If you lean back a little bit, you're looking at a sort of trapezoidal bunch of stars (including the sixth-brightest in the sky, Capella) that make up the constellation Auriga, the charioteer. Lean dizzily further back and you're looking upside down at the North Star. You might have more common sense than me, and just turn around to see it, Polaris, at the end of the Little Dipper's handle.

But in deep winter, I stop having that kind of common sense. The kaleidoscopic night-shining sky rattles my brain. I wonder if there are aliens out there.

This question by itself is pretty commonly asked nowadays, so there's no real lack of psychological balance there. It's the possible answers that bend your interiors out of shape.

The gargantuan number of stars in the universe implies there's a reasonably good chance creatures of advanced intelligence exist elsewhere. This is a commonly accepted basic line of reasoning. But if so, asked Enrico Fermi (a superhero of twentieth century physics) back in 1950, why haven't we seen them? This is known in UFO and search for extraterrestrial intelligence studies as the Fermi Paradox.

Seven decades later the evidence for the possibility of extraterrestrial life has increased. As of December 2020, astronomers had confirmed the existence of more than four thousand planets orbiting other stars. A study in 2013 reported it's likely that almost all of the hundred billion or so stars in our galaxy have planets revolving around them and that 17 percent of those stars probably have a roughly Earth-sized planet in a tight orbit. Estimates of the total number of galaxies in the universe are as high as two hundred twenty-five billion, and some astronomers think that's probably a conservative figure.

So with all these possible homes, where are the putative aliens? The journey from star to star is prohibitively long, so maybe there's no sign of them because they just haven't gotten here yet.

Is it really true that there's no sign of them?

It depends on how you read the thousands of UFO reports that have been investigated and confirmed to be neither hoaxes nor mistakes, but actually unidentifiable objects. The fact is that UFOs and the phenomena associated with them are very well-studied by credible, credentialed researchers and by government agencies such as the CIA, National Security Agency, U.S. Air Force, and more.

And then there are the reports of people being abducted by nonhuman entities. These incidents, too, undergo meticulous investigations. In the 1990s, Harvard University psychiatrist John Mack spent several years interviewing scores of people who described abduction experiences and analyzing their descriptions. He concluded that neither mental illness nor dishonesty was a prevailing problem among them, that there is no conventional scientific explanation for whatever is happening, and that the people's own stories best describe it.

A knowledgeable correspondent explained to me recently that "it's fairly clear that we are being visited by multiple species. ... I think some of the visitors are interplanetary, some are inter-dimensional and some are time-travelers."

Like Mack's study, this is not dreamy science fantasy. These categories of alien origins are more or less conventional in serious discussions of abduction narratives.

Stephen Hawking, of all people, warned that SETI programs are inherently dangerous: the aliens, should they arrive, may have bad intentions and the capability of carrying them out. *So do we really want to attract their attention?*, he asked not too long before his death in 2018, turning the Fermi Paradox into the Hawking Admonition, perhaps.

I guess my credibility as a backyard naturalist is probably shot when I disclose that the winter sky prompts such thoughts. But winter is prohibitively long here, it's just getting under way, and my mind has miles to go when the snow's too deep for the rest of me to get very far afoot. If there's anything the last five hundred years of science has taught us, it's that the universe keeps turning out to be much, much larger than we thought. You really don't get to pick.

The Dark Universe

When you step onto your porch late at night and stand there with your hands in your pockets looking up, the cosmos seems to be teeming with light beings. Your eye can pick about six thousand individual stars in a light-pollution-free sky. Enough to keep a backyard naturalist occupied for life. Behind them are the sprays of the Milky Way, which, if you go back in and get a pair of binoculars, themselves turn out to be made of stars.

These are just the stars visible from Earth in our galaxy, which altogether contains a hundred billion stars, or more—different astronomers argue for different estimates. A few billion here, a few billion there, pretty soon it adds up to some big galaxy. A study in the 2010s estimated there are about ten times more galaxies in the universe than was previously thought. Instead of the previously estimated around two hundred twenty-five billion galaxies, the new estimate is upwards of two trillion.

A lot of stars. An average galaxy, one of the authors of the study estimates, contains around a hundred million stars. A hundred million times two trillion is a mind-derailing number. To say the universe is immense would be, well, the understatement of the week.

Now try this: the two trillion galaxies are not the whole universe. It's bigger. A lot bigger. In fact there is good reason to believe that all the galaxies and all their detectable stars, planets, gas, dust, protons, neutrons, electrons—and living beings—make up only about 4 percent of everything that exists. Ninety six percent of everything that exists has never been detected.

What is this unseen 96 percent? The short answer: no one knows.

The long answer is, of course, mind-derailing.

The whole universe is expanding, which on the large scale means that all the galaxies are flying away from each other. A way to picture this is to think of a balloon with circles drawn on it. As you blow up the balloon, the circles don't go anywhere, but they get farther apart. That's roughly what's been happening in the universe since it exploded into existence about 13.7 billion years ago at the Big Bang.

Now, gravity develops when a mass creates a bend in space-time. Masses can be measured by observing a planet's or star's motion with respect to other objects. For example, the Earth moves around the sun at about sixty thousand miles per hour. If the sun were four times more massive, the Earth would move at twice the speed. The same equations work for the stars circling around the center of the galaxy. By measuring the stars' speed, the astronomers can estimate the mass of the galactic center. This relationship of mass to motion works throughout the universe, as far as anyone knows. This obviously oversimplifies things, but hopefully you get the picture.

Galaxies, being enormously massive, create enormous gravity. The two trillion galaxies all together have so much mass that their gravity should be pulling everything back and slowing down the expansion of the universe. The trouble is, other reliable measurements show the expansion is not slowing—in fact, it's accelerating. How could this be?

The answer—many cosmologists think—is that an enormous amount of undetected matter and energy is exerting a force counter to the force of gravity. This unseen matter and energy does not interact with electromagnetic energy, or light, so it's referred to as "dark." Dark matter appears to make up 24 percent of the universe, dark energy not quite 72 percent, and the "baryonic" matter familiar to us, about 4 percent.

What is this dark matter and energy? It could be in the form of colossal numbers of brown dwarfs, which are small stars too cold and dim to detect. Or it could be in the form of supermassive black holes, so dense that even light does not escape them. Or it could be in the form of particles and forces completely different from the matter known to us; since this "non-baryonic" matter doesn't interact with electromagnetic energy, we don't see it.

Ninety-six percent of everything that exists is undetectable to our five senses.

As far as I know, the metaphysics of this astounding fact has not been explored very far because there's almost nothing concrete—or abstract—to go on. It's there. It's greater than the six thousand stars in your sky and all the distant galaxies in our swiftly tilting cosmos. Be careful how you think about this. You can get derailed.

Winter Moons

"The spruces rough in the distant glitter
Of the January sun"
   —Wallace Stevens, "The Snow Man"

Here in Maine, winter is long and cold. In the past it was even longer and colder, or so memory and certain old people suggest. One of Thoreau's journal entries from before 1850 notes without surprise a fairly heavy snowfall in mid-April. That was in Massachusetts, which is still part of the temperate east. Farther north and east, beyond New Hampshire, the Saco River in southern Maine is the accepted anthropological divisor of the Eastern Woodlands region from the Eastern Subarctic region. "Subarctic" refers to a length and depth of winter freeze which is something less than polar. The limits of a subarctic winter might accurately be described as the months when snow can reasonably be expected to bury everything. And in earlier times, the months when food became scarce to nonexistent. Most of Maine drowses through winter from November to April.

By late January the cold has normally been so long and so thorough that it's difficult even to stay awake. The ice grinds everything to a halt, or so close to a halt that for all intents and purposes nothing happens. The roads are frozen, and so are your bones. Snow piles up everywhere, obliterating the driveway and the baseball diamond. In fields the only signs that anything ever lived are dead spokes of grass or a few uncut corn stalks, the occasional raccoon, fox, or deer tracks in the snow. You don't want to go outside. You want to stay in where the heat is, smell the wood smoke or the dry, nylon odor of electric coils. The inner staleness of the kitchen, unventilated since October.[1] A house is a core of warmth, like a burrow. It seems unutterably small after a while, but at least it's not frozen. In the cold we describe even the warm by what it's not.

Cold is the absence of heat. Ice is that pervasive presence in the universe that signifies what is not. Sometimes on really arctic nights the ice—or maybe not the ice itself but its stillness and hardness—becomes fascinating, and I feel sucked outside to see the emptiness. Away from the artificial fires of western culture, which throw smoke and black soot all over chunky roadside snowbanks, the snow in the woods remains purely white, even during its porous melting period in March. The whiteness is a blankness, even more complete than on the ocean surface because it does not move. Billions of tiny frozen water crystals, motionless, piling up around the hemlocks and in the arms of pines, bluish in moon shadows. Everything suspended, waiting for the sun to come up.

Especially at night. The air is emptied of moisture, and to breathe is to suck in pure cold, like blocks of ice tumbling into your lungs. In arctic cold, minus 10, minus 20 Fahrenheit and colder, a deep breath extinguishes the vascular heat in your chest, and a sharp pain creases your sternum. You breathe slowly to preserve the inner reserves of warmth.

I saw the emptiness completely one moonless January midnight when I walked across the pond to look at Orion. The camp road was slick with crushed powdery snow over a slab

---

* This essay was written in 1989 and first published in 1996. It portrays what winter consistently used to be like, and retains here most of its original phrasing. But if it were written in 2020, descriptions and phrases such as "unventilated since October" would appear as "unventilated since November," in accord with climatic changes unofficially observed over the past thirty years.

of ice. The stars were thick, like magnified crystals in the blackness. On the pond my boots blasted oblong impact pits into the glazed snow. I thought the pond must be frozen completely through to the bottom.

Everything seemed impregnable, as if the cold itself was insulation. In the first stages of freezing there is nervousness. When the chill penetrates your skin, you have a natural inclination to move, which for most people means shivering. As the cold filters further into your bones your body becomes calmer, and drowsiness takes over. A desire to succumb sets in, like a cat settling into a chair, and a fascination for sleep dulls the desire to survive. In its final phase, I imagine, it solidifies into a need to relinquish consciousness completely and become ice. Standing on the pond, binoculars in glove, I kept shivering. The emptiness yawned all around me. Flat, dark ice reposed like a moonscape, sometimes buckling and creaking as if the Earth itself could shiver. In a rough circle around the pond's edges loomed pointed giants, spruces and pines.

It was like standing in a still crater. Rim mountains spoked up all around me. The impact basin was flat, pocked with tiny holes. The arctic cold of the Earth, I thought, is the same as the moon's, or Triton's, or Charon's. Absence is absence. Nothing is nothing. You can die of sleep as easily here as there. For a few minutes I relaxed. Stars plainly rising over a crater-rim scintillated on the edge of the absence, like the fat dreams that come before deep sleep. I was on a moon somewhere, becoming ice.

Without closing my eyes I passed for seconds or an hour outside the normal orbit of perception, under the snow-encrusted surface of the pond to the stillest, blankest observable territory in these parts. I slid across an empty, godforsaken whiteness and fortunately for my soul, recognized Europa circling Jupiter half a billion miles away. A place of uncontrol-lable cold, smooth beyond comprehension because it's wrapped in a frozen ocean, solid ice fifty miles or more straight down. Underneath is more water, not yet frozen, rest-ing on dense, cold rock. For nearly eight thousand miles in a circle around Europa's equator there is only ice, unrelieved. Any hills or ridges slope very gradually to a few hundred feet at most. Europa is smoother and rounder than a billiard ball and stiller than an arctic lake.

I flexed my fingers in my gloves. This is winter. Beyond the mania of its three-and-a-half-day orbit, Europa's only ac-tivity is the creation of more ice. When its surface buckles and cracks, more water and rock fill the fissure and freeze. No one knows exactly how this happens. You might hear a crack like thunder explode the silence, and the ice might jerk and split at sonic speeds beneath your feet, and pressurized water flood suddenly into the crack. Or more likely, the ice warps as it does on the pond, with creaks and twists and throatless barks. Water then seeps in slowly from underneath and freezes in the minus 270 F cold. Craters are erased by this pro-cess in short periods of time, say thirty million years.

Otherwise there are no shapes or points of reference on Europa, no figures at all. Underneath the ice enormous lines criss-cross like Martian canals, but to someone standing on the surface they'd be invisible and wouldn't exist. There would be nothing to see or hear for millennia at a time, only the disk of Jupiter, four hundred twenty thousand miles away, filling a sixth of the sky, illuminating flat, unending, polished nothing.

Survival here would involve the generation of motion in utter stillness. The excitation of molecules somewhere, at some innermost area which can hold out the cold. The real activity of winter is whatever supersedes the absence of warmth. On Europa the problem of shelter would concern not only the heat of your physical body but, the same as in Maine, your mind, which would be cooped in by the cold itself, not just the bad or absent roads. On Europa you would invent shapes and figures merely to survive. Your mind would seize upon a gloss of planetlit ice and thaw it into someplace else.

To be clear: heat is motion. Temperature is a measure of molecular activity. Absolute zero is the total absence of heat energy, the temperature at which all molecular motion stops. This condition is abstractly represented on the Kelvin scale by the expression 0 K. On the Celsius scale it is minus 273. Fahrenheit, minus 459. It's difficult to imagine that kind of cold even on Europa, which is relatively warm at minus 270 F, or in Maine, which is relatively warmer, though not much.

The cold is not the problem because you can always give in to it. You can always sleep, drift away toward death, when your energy subsides. You can extinguish much sooner than absolute zero; human beings seem to succumb to the stillness at levels of inactivity much more active than even Europa's. In arctic Canada people do not go out alone when it's colder than minus 50 F. The core of what's human can solidify long before hell freezes over.

Even more difficult to imagine in deep cold is the survival of consciousness. On the surface of Europa, as on the frozen pond, consciousness would be simply the search for warmth, which, again, is activity. The moon-blankness is startling in a way, like the blankness of the snow-encrusted pond. The mind shivers at the thought of Europa. It's so cold, it's bright: it reflects sunlight and Jupiter-light with an albedo (or reflectivity) of .62, in astronomy-speak. The Earth, blue and white in space, has for comparison an albedo of .39. The moon's is .11. Europa's surface temperature is on average lower than its neighbors' because it actually reflects heat away. Of the four Galilean moons of Jupiter, Europa is (with Io) in a warmer orbital zone than Ganymede and Callisto. In the strain and tug of Jupiter's gravitation, Europa is geologically active deep inside, pushing ice and rock slowly upward in a way that Ganymede and Callisto don't. Io is a different story. It is roughly as cold as Europa at the surface, but a place of volcanoes and sulfur compounds, a burning upper hell in contrast to Europa-Cocytus, lower hell. Triton, circling Neptune billions of miles farther on, is the coldest place in the solar system, with a mean temperature of minus 391 F. But even there, seasons lasting decades alter the surface ice, and eruptions of nitrogen crystals blow through the tenuous atmosphere.

By contrast, Callisto, seven hundred fifty thousand miles beyond Europa, is completely motionless. Its surface temperature is about minus 253 F, but nothing internal disturbs that surface, at any interval of time. Rocks inhabit its ice, so it's almost as dark as the moon, albedo .17. Callisto's cold, unlike Europa's, takes form. Instead of an absence of shapes and figures, on Callisto things rise up like monsters. Jagged craters cover everything. Most are small, less than ninety miles across, compared to those of other planetary bodies. Most are craters cratered with craters. Astronomers weirdly call these geographic features "palimpsests," as though the craters were trying to tell us something. One impact basin called Val Halla appears to be an enormous exception in size, with concentric rings emanating outward more than a thousand miles. But it too is webbed with smaller craters and has no ring mountains like those on the Moon.

The cliffs and small mountains of Callisto are older than Ganymede's and more motionless than Europa's ice. They loom like the monstrous shadows of frost. They seem to be the cold itself, frozen into shapes whose presence intensifies the absence. Apart from its inert, perpetual orbit of Jupiter, nothing is happening on Callisto. The meteorites that billions of years ago blasted out the craters lie in pieces the same as the night they hit. These are hell places. Whatever consciousness can survive here consists in isolated hills and shadows, and the disturbing fantasy that even Jupiter, which at some remote time might itself have shone like a star a thousandth as bright as the sun, has ceased to pay attention. Its glimmering night-light passively reflects from Callisto's dirty, snowy surface. The dark mountain points and frozen impact escarpments form the shadows of abandonment. Even the sun is too distant to move anything here.

Those frozen mountains are horrible. They are what you would see if you could somehow see nothing. Europa in its smoothness is in a sense invisible, rejecting everything, every attempt to locate anything. It throws all activity of heat or mind directly into empty space. Callisto outlines what a hole must be, sunken deep toward absolute zero. Except for its rock and ice, it does not, for all intents and purposes, exist.

* * *

Space contains our parallel worlds.

A winter night in Maine, like Callisto-night, seems eternal. The jagged spruce-tops in the woods beside the frozen lake point toward burning white crystals, like the mountaintops of Callisto. As I looked up that January night, the darkness seemed like an envelope sealed for good. Daylight would never arrive, warmth would never return. The entire universe for a few moments in the cold was eternally absent.

Winter is night and cold dissolved together. It is the living figure of all absence, and it brings all conditions of absence with it. It came to me that the Abenakis and the Red Paint people, natives of Maine centuries and millennia before us Europeans, must have understood winter in ways impossible for me. They didn't think of bald Europas or Callistos, but instead moved with the season, entered it, maybe. When the cold had halted everything serviceable to life, the Abenakis looked into the same sky as I do and called the bright full wafer there Starvation Moon. The clams and mussels that were dried during summer had been eaten, and the deer and moose had all but disappeared. The bears slept, and nothing grew or even moved except in creaking wind and snow. Hunger, they knew from firsthand information, is a condition of nature. It is the living experience of nothing. It could not be defeated because it does not explicitly even exist, except in a figure of the changeable moon. It could, given the right reserves of caloric and psychic energy, be waited out, the way the drowsiness before oblivion can be blunted.

I, on the other hand, shivering in the snow, knew the wood stove was warming the house. We have ways of preserving bodily heat during winter. But even nowadays in Maine, "cabin fever" is not an altogether figurative expression. Some people fail in winter. The cold grinds everything to a white smoothness. It slows the elbow joints and knees, and seems to freeze the afternoon itself, and evening, to a standstill. The sun is absent fifteen hours a day here in the dead of winter, the Earth tipped over on its axis out of heat and light. Night enfolds the world by four o'clock. Feeling sealed off, like having claustrophobia in the darkness, people fray, their normal reserve and tolerance dry up. A winter nervousness sets in that sometimes ends up bad, in fits of rage, or brawls, or worse, for all the world like psychic shivering. In winter we sleep at the edge of oblivion. The place of hell itself. The tops of spruce against the sky become ancient, inert mountaintops.

In times before the Abenakis inhabited the Eastern Sub-arctic, a seed philosopher for modern Europe, Parmenides, pictured two ways of understanding the universe. One way was to think or realize that things actually do exist. To imagine for example Callisto and the moon out there, with frost mountains. And millennia later to imagine, on one at least, latticed footprints undisturbed for the last twenty-six years and for twenty-six thousand more. The other way was to study or assume their emptiness. If we rationalize intently about Europa and Callisto for a time, we infer and—right on the edge of possibility, or its failure—almost perceive that they are essentially void, essentially nothing.

But no real information comes back from this way. Focus your mind on absence and things begin to freeze together and become indistinguishable. Pieces of facts, like inert meteor debris strewn around an ancient crater, can be heaped together, but the reality of Callisto or Europa, from this view, is their essential vacancy. You end up chattering about nothing. It's like shouting into an empty chasm and wishing for a rock slide. It's a wish for annihilation, like surrendering to the cold for a quick rest, and sleeping. An Abenaki under the Starvation Moon did not rationalize the absence of meat or the emptiness of her belly. To visualize absence as though it were present reveals eventually the absence of absence, things exactly as they're not. Thoughts of what is not are thoughts of annihilation. Thoughts of annihilation are thoughts of hell.

The problem of winter is to generate activity, to excite molecules to warmth. The cold wants motion, existence of some kind, even a dream. Warmth is generated not by, but among the bare black crater-rims and frost. Not the motion gulped in by the eyes, which at the right depth will freeze over, but the motion generated in desire, which outlives starvation. The pure experience of cold is to inhabit it. Trees purify themselves by dropping leaves and draining sap, and enter January bodily. They neither succumb nor disappear, but occupy winter. There is no more ancient activity in the universe than the seeds of heat inside the cold.

The mind aligns the emptiness. It frames the snow and ice, and zero temperatures, and spruces which are little more than crater-edges. Black, empty space becomes eternally active. Seven crystals magnified in winter air are sisters, Pleiades. Three others are a hunter's belt. Seven in the north become a massive whirling bear. Europa's ice and Callisto's mountains are gateways into hell. Snow is a natural condition of April. Bark in the Adirondacks becomes food. The simple act of waking on a winter morning when the fire is out requires a mind of winter, seeing nothing that is not there, and the nothing that is. And minutes later, out of empty black space comes sunrise.

Europa

About twenty-five thousand years ago, our part of the world was covered by ice a mile and a half thick. It extended a hundred and eighty miles out beyond the rock and sand we now call the coast of Maine. In the next thousand years the Earth started warming up and the ice slowly melted. By about fourteen thousand years ago, ocean shoreline was exposed at the vicinity of Farmington today.

The point here is not that winter sometimes seems cold enough to maintain a glacier, but that underneath the ice there was not only rock, but everything we know and love today. In particular, ocean, where we all came from. The astronomers believe that hidden oceans of liquid water might not be unique to paleolithic Earth. Several other floating orbs may have internal oceans, including Pluto, of all places, and Saturn's moon Titan, and Jupiter's moons Callisto, Ganymede, and in particular Europa, which is completely covered with ice.

It's kind of unsettling to think about. Europa circles silently around and around in Jupiter's magnetic- and gravity-ridden space. It's white and smooth. The ridges, pits, cracks, and grooves in its ice rise and descend no more than a few hundred yards, which is about the terrain you'd find if you expanded a billiard ball to just less than the size of our own moon.

Europa also has a thin atmosphere of mostly oxygen, odd for a moon, or anywhere actually. It's thought to be formed from sunlight striking water molecules and splitting them into their two atoms of hydrogen, which is light and drifts away, and oxygen, which is heavier and stays. The unusualness of this is not the gases as much as it is the water that gives rise to them: Europa's ice shell is estimated to be fifty to one hundred miles deep, and underneath that is believed to be a saltwater ocean. The Hubble Space Telescope at one point spotted what appear to be eruptions of water vapor from the ice, corroborating the theory.

How could there be an ocean in a place whose average surface temperature is minus 260 degrees Fahrenheit? Well, the warmth that keeps the water liquid is generated through "tidal heating." The whole moon is twisted and flexed by the interacting gravity fields of Jupiter (whose gravitation is so big it keeps Saturn in line) and its three other large moons. The twisting creates heat deep inside. Evidence of the heat are flows of ice on Europa's surface, not liquid rivers but ice that emerges from inside which is warmer than the dead-cold surface.

There is a lot of scientific excitement about what that warmth might imply about Europa's underworld. If its inner ocean is liquid, then it's reasonable to wonder if life could have sparked and evolved in there. How anything could live in total darkness under ice fifty miles thick, no one is sure. But "extremophiles"—creatures living in environments fatal to the rest of us—have been found in very unlikely spots here on Earth, like in near-boiling water, in radiation that would kill a human, and in ice. Some years ago, NASA researchers drilled a hole in the Antarctic ice sheet and stuck a video camera down there to see what was going on with the water, when a three-inch-long shrimplike creature muckled onto the camera. It lives there under the ice. If things can evolve so they live under Antarctic ice, then living things might be evolving inside Europa.

These are just cold, hard facts the astronomers give. Cold up to the point, anyway, where the possibility of something living inside Europa twists up your imagination and different kinds of fact-based excitement start bubbling into thoughts.

Wandering reflections on the possibilities of extraterrestrial life are fascinating, but can get kind of unsettling. Before he died, Stephen Hawking warned that searching for intelligent extraterrestrial beings is risky because if they exist, they might not be friendly. So far, nobody thinks that any possible Europan beings are intelligent in our way of defining the word. If they exist, they're more like those shrimpisms, or much smaller. Or maybe like the monster in the recent, detail-accurate film "Europa Report."

But really, how do we have any clue what's evolving under the ice? What unthinkable things right now, this exact moment, are twisting, furling, and muckling in that frigid blackness?

Anyway, these are just thoughts wandering off and twisting around in the winter dark. We don't know what's under that Europa-wide glacier, waiting to be exposed.

Pluto Borderlands

The New Horizons spacecraft is on a quest for knowledge. It's a robot about the size of a compact car and weighing about a thousand pounds. It uses two hundred twenty-eight watts of electricity generated by a radioisotope thermoelectric generator fueled by twenty-four pounds of plutonium-238 oxide pellets and has an array of imaging, radio, and particle and spectrum detection equipment.

Its roughly fifteen-year mission is to fly close to Pluto and its moons and then into the solar system's borderlands, gathering data. It took off from Cape Canaveral, Florida, on January 19, 2006 (the one hundred ninety-seventh anniversary of Edgar Allan Poe's birth), and about a year later bent around Jupiter where it took some pictures and got a slingshotlike boost that increased its speed to about fifty-two thousand miles an hour. After that it was more or less a straight shot to Pluto, about three billion miles from planet Earth. It arrived July 14, 2015, and kept chugging into the Kuiper Belt.

It took space scientists nearly two decades to get the mission off the ground because the people who dole out the money for such projects weren't really sure what was to be gained (or who would gain it) by flying to Pluto. It is, after all, far away, and what kind of wealth does an expensive mission like that promise? Even the information you got back would be destined mostly to oblivion in high-powered university computers and esoteric graduate seminars in planetary astronomy. But after a lot of back and forth which included a healthy dose of simple public curiosity, the U.S. Congress in 2002 approved funding for a scaled-down version of the original, early-1990s project proposals.

The mission has three main objectives: to accurately detail the perpetual winterworld geology and physical structures of Pluto and its moon Charon; to detail and map their surface chemistries; and to detail Pluto's atmosphere. Seems pretty simple. Of course it's not, because so many particles of information are radioing back to Earth that the scientific analysis and conclusions process will take decades. At least.

* * *

The physical cosmos is mind-bogglingly detailed. Whole worlds that you can't see are crawling around on your coffee cup rim while you read this, and they all have intricate patterns, structures, chemistries, and interrelationships that took at least 3½ billion years to evolve and have so far taken hundreds of years to track down and describe. The work has barely begun. Even on the next larger scale up from us, meaning the moons and planets, the detail is so fine that just spotting Pluto, which is one thousand four hundred thirty or so miles wide, took decades of hypothesis and observation.

People started speculating seriously in the latter part of the nineteenth century that there might be a planet beyond Uranus (first identified in 1781) and Neptune (tracked down in 1846), and various people, including Percival Lowell of the Martian civilizations theory, invested copious mathematical energy trying to predict where and how large it might be. On the momentum of Lowell's efforts (he died in 1916), the young astronomer Clyde Tombaugh in 1930 found a speck of light changing places from one spot to another on photographic plates, which turned out to be Pluto.

Just a speck of light out there, indistinguishable from millions of other specks unless somebody with a telescope guides you to its exact location and brightness. So far away and faint with reflected sunlight that even the Hubble Space Telescope, the greatest device of its kind so far, gets only shadowy images of it.

Curiosity about shadowy, seemingly distant things is always radiating in somebody's mind, including the astronomers who lobbied for the New Horizons mission, the mathematicians who wondered if Pluto was there before it was there, and generations of philosophers and mystics whose sense of the cosmic shadows indicated that (for example) "there are innumerable suns, and an infinite number of earths revolve around those suns, just as the seven [planets] we can observe revolve around this sun which is close to us." How Giordano Bruno knew this, or guessed it, decades before Galileo trained a Walmart-quality telescope on Jupiter and its moons, is not clear, but it turned out he was essentially right.

As if the microbes, Earth, and solar system were not detail enough to last an inquisitive species a lifetime, the universe is so vast that there are chips, stones, and grains of dust lying on the surfaces of planets orbiting stars in mind-boggling numbers. Some astronomers think there could be ten quintillion (10^{19}, or 10,000,000,000,000,000,000) planets in the Milky Way galaxy alone. Some guess there could be more than a hundred billion galaxies—as many as two trillion.

From this view, Pluto really is a speck of dust. How could the details of its geology, chemistry, and atmosphere be so much more important than any other grain in the cosmos that a mighty concentration of human mental energy gets set in motion and crystallizes in a half-ton of carefully configured metals, electricity, and radioactive isotopes to get sent off specifically to find them out?

Well, one answer is: they're not that important. But another answer is: they are far away, and shadowy, and reachable.

The scientists put almost all their mental energy into collecting and categorizing physical facts. But occasionally, when nudged out of their data dreams, they offer a superficial cliché or two they hope will characterize the original impulses that drew them there. Barrie Jones in his book *Pluto: Sentinel of the Outer Solar System* states it like this: "Aside from [the] scientific justification [for sending a robot to Pluto] is the human drive to explore, a drive that permeates almost all cultures, and has done so throughout human history." This is a rehearsal of George Mallory's more concise expression of his reason for climbing Mount Everest: "Because it's there."

By and large the scientists do not say much more than this about it because their interests, after all, lie in the physical facts, not the metaphysical motivations. But a conversation has been going on for a hundred years or more—largely in the academic shadows, as it were—about what effect the scientific disposition toward what constitutes useful knowledge has on our general well-being. The hyper-rational philosopher Alfred North Whitehead warned in the early decades of the twentieth century of the limitations of hyper-rational thinking. The astronomer Arthur Eddington around the same time pointed out that, as powerful and practical as scientific inquiry is, there are nonetheless parts of reality that science cannot treat. Why is a joke funny? Scientific analysis can't explain it, and annihilates the humor in the effort.

There are other kinds of knowledge besides scientific. We know this. There is a disposition in science, though, that does not want to talk about other kinds of knowledge and views them as either beside the point or simply illusory. The nearest many scientists and philosophers of science come to talking about other kinds of knowledge appears in phrases such as "the human drive to explore" — we explore because we explore.

Even detailed explanations of what phrases such as "the human drive to explore" mean do not seem well-mapped to many scientists. Such explanations depart the world of physical facts too quickly and enter those far-flung parts of reality that science has trouble treating. Giorgio de Santillana and Hertha von Dechend call one of those territories "mythical knowledge." They argue in *Hamlet's Mill* that ancient cultures processed and preserved information about the stars, planets, and their motions through the telling of stories, symbolic rituals, and music. A human being living in such a culture would "have been a participant in the process of mythical knowledge ... In his own person, he would have been part of a genuine theory of cosmology, one he had absorbed by heart, that was responsive to his emotions, and one that could act on his aspirations and dreams. This kind of participation in ultimate things, now extremely difficult for anyone who has not graduated in astrophysics, was then possible to some degree for everyone, and nowhere could it be vulgarized."

In other words, knowledge of the cosmos was not simply a grasp of physical facts and their explanations, but a participation in them through "an idea of the overall texture of the cosmos." That texture is not a rational explanation of physical forces and interactions happening around and to you, but an experience of your place and its activities fitting together with your own mind.

Other scientific-age thinkers who tried to map the experiential territories that seem beyond the borders of scientific inquiry include for example Carl Jung and Joseph Campbell. But despite these forays, we maintain a general cultural assumption that scientific inquiry and analysis is the most highly evolved mental activity human beings can engage in. If prescientific peoples had their own mythologies, those are sort of interesting and quaint and valid *for them*. But *for us* here in 2015, knowledge progresses evolutionarily and rational science is both the pinnacle and the cutting edge of that progression. The mind-boggling beneficences of medicine, communication, and transportation that result from applications of scientific thinking to physical details seem to bear this out. On that momentum, the robot heads to Pluto.

* * *

Out there where New Horizons is foraging for facts, it is unutterably vast and lonely. Pluto is on the edge of the solar system, meaning that on average it's something over thirty-nine Earth-sun distances (or astronomical units) from the center. Another way of stating its distance and isolation is to observe that, as telescopes got better during the 1980s, '90s, and 2000s, the astronomers grew less sure about what Pluto actually is. It started out as the ninth planet in the 1930s but by 2006 got recategorized—amid considerable dissent—to "dwarf planet" because it turns out that at least one thing out there is bigger than Pluto, and probably more. For the solar system has, in one way of describing it, three parts.

First, there's the solar system of conventional description. This includes the eight major planets (inner planets Mercury, Venus, Earth, and Mars, and outer planets Jupiter, Saturn, Uranus, and Neptune) and their moons. It also includes the asteroids, which are sort of miniature planets and planet chunks orbiting the sun mostly between Mars and Jupiter, but also other areas. In the vicinity of the outer planets they're called centaurs.

Second, just beyond the orbit of Neptune is a borderland called trans-Neptunian space or the Kuiper Belt, or Edge-worth-Kuiper Belt, depending on who's getting credit for predicting its existence. The Kuiper Belt is made up of asteroidlike chunks of rock and ice, collectively called Kuiper Belt Objects, of varying sizes orbiting between about thirty and fifty astronomical units from the sun. Pluto and slightly wider and more massive Eris are the two largest known KBOs among an estimated seventy thousand with diameters of more than about sixty miles. The most distant large KBO detected by 2015 is a dwarf planet called, as of April 2014: 2012 VP113. Its nearest approach to the sun is about eighty astronomical units, and the farthest extent of its orbit is estimated to be about four hundred forty-six astronomical units. It's a little smaller than Sedna, whose closest approach to the sun is about seventy-six astronomical units, and at farthest is around nine hundred thirty-seven astronomical units away. It takes 2012 VP113 about four thousand two hundred seventy-four years to circle the sun once, and Sedna more than eleven thousand years.

Third, beyond the Kuiper Belt and Sedna is the Oort Cloud, where an unknown abundance of comets slow-orbit the sun. No one has ever seen anything in the Oort Cloud because whatever's out there is too small and distant for telescopes—one thousand to one hundred thousand astronomical units out. But the Oort Cloud is thought to be the origin of long-period comets, meaning the ones that come streaking into our view just once, to return near Earth only after thousands of years, or possibly never.

The farther you get from the sun, the more frayed everything is. Mercury, Venus, Earth, and Mars occupy orbits relatively compact to the central fire. A giant leap out from Mars, which is on average about 1½ times as far from the sun as the

Earth, the next big thing is Jupiter, looming so huge, there, 5.2 astronomical units from the sun, that its gravity acts like a shepherd on everything inside the Kuiper Belt. Early on in the formation of our neighborhood, Jupiter's mass (equivalent to about three hundred nineteen Earths) was so mighty that it pushed Saturn, Uranus, and Neptune into the stable orbits they inhabit now. Saturn, more than three times smaller by mass than Jupiter but still enormous, in turn perturbed Uranus and Neptune, and those two, about seven times and five times smaller by mass than Saturn (but fourteen and seventeen times larger than Earth), swept their neighborhoods clear of many rocky, icy items like or smaller than Pluto, and those are the Kuiper Belt Objects and comets. Circling around the sun like spray on the edge of everything.

* * *

After it cruised past Jupiter in February 2007, New Horizons went into a hibernation mode with a few wake-up periods to conduct checks. In the next eight years, the empty space it traversed was, for all practical purposes, empty. Uranus and Neptune tooled along in other, far-flung parts of the solar system. The robot is not equipped to detect the scant sort of details lurking in that part of interplanetary space.

About two hundred days before its closest approach to Pluto, New Horizons grips down, begins to awaken, and starts taking pictures and measurements. In mid-July 2015, it sweeps first within about six thousand two hundred miles of Pluto's surface, and fourteen minutes later within about sixteen thousand eight hundred miles of Pluto's large moon, Charon. Radio signals pour data toward Earth. They take more than four hours to traverse three billion miles.

The astronomers were fairly confident about some of the details New Horizons would find. Pluto's surface is covered by nitrogen ice, which if you were to walk on it would have a granularity like snow. There are traces of methane and carbon monoxide ices, and just under the surface of craters and otherwise rocky debris there is likely water ice. The astronomers eagerly anticipate the presence of water anywhere because it implies the presence of processes familiar to Earth. There is a sense, when water is detected or predicted in any quantity, that life is possible—either in the distant past, as on Mars, or in the present, as underneath Jupiter's moon Europa. No one predicts anything alive on Pluto.

The ices on Pluto's surface vary with its seasons. Like all the planets, Pluto's orbit is elliptical, and also, like Earth in particular, its axis of rotation is tilted with respect to the sun. These two factors combine to create changing conditions of light and temperature on the surface of both. But on Pluto the effects are extreme. While Earth's tilt is 23.5 degrees, Pluto's tilt is 57.5 degrees, and while Earth's orbital eccentricity (meaning how much the orbit's elliptical shape varies from an exact circle) is 0.0167, Pluto's eccentricity is 0.251. These extreme variabilities, together with the fact that Pluto is so far from the sun, create distinctly different conditions of light and temperature—different seasons—in Pluto's two-hundred-forty-nine-year year.

Pluto's northern and southern hemispheres tip gradually into and out of the sun's direct rays just like on Earth, but with steeper shadows at the winter solstice and more direct light on the poles at the summer solstice. The speed of any planet (or dwarf planet) varies according to its location in its elliptical orbit, faster when nearer the sun and slowing down when farther out. Because of its steep inclination and its variable orbital speeds, Pluto's seasons are not of equal lengths, as they are (roughly) on Earth.

Summer and fall in Pluto's northern hemisphere last about forty-two Earth-years, while winter and spring are about eighty-three Earth-years long. The southern hemisphere, meanwhile, has a shorter spring because Pluto is closer to the sun during that spell and therefore traveling faster. Because it's so much closer to the sun during its spring, its temperatures then are probably higher than its temperatures in summer. Its thin atmosphere is thought to expand in summer and shrink in winter in response to warmer, sunnier and colder, darker conditions.

The astronomers treated most of this as facts before New Horizons' encounter.

* * *

The scientists are essentially, pragmatically uninterested in how these facts about ice, sunlight, winter, and summer on Pluto fit your mind. But to wander off the beaten way of science for a moment, there is something familiar to someone who lives in, say, Maine about the idea that winter and spring could be nearly twice as long as summer and fall. In the hinterlands of the imagination in the Great White Northeast is the perennial feeling that the stretch of time from November to about April lasts, well, about eighty-three years, while the stretch from May to October is surely half that, or less. In Maine's deep winter, snow and ice cover everything. The air dries up, the nostrils and throat are perpetually irritated, and human skin turns to ancient papyrus.

On Pluto, similarly, there is more nitrogen snow on the ground in winter because of something similar to dryness, and less on the ground in summer's warmer temperatures because some of the snow sublimates into the atmosphere. Pluto's northern hemisphere sees its next winter solstice in the year 2029. Its autumnal equinox was in 1987. On a January afternoon, the beauty at Maine's autumnal equinox seems decades past.

Of course, the average temperature on Pluto's surface is around minus 370 degrees Fahrenheit. Even the bitterest January day in Maine is not quite that cold. Normally. And while in some southern backyard views in Maine the winter sun barely clears the fir tops, at similar latitudes on Pluto it descends more than twice as low because of that 57.5 degree tilt. The shadows there are twice as long and dimmer.

And then there is the matter of Pluto's distance from the sun. The Earth is one astronomical unit— about ninety-three million miles—from the sun. Pluto is on average about 39.6 astronomical units from the sun. In Pluto's black sky, with atmosphere too thin to scatter blue light the way Earth's does, the sun shines eight hundred times less brightly than on Earth. It's a sort of perpetual twilight, somewhat brighter than a full moon, just enough light to read a book. But that landscape of frost and craters, mountains, cracks, and possibly ice geysers, would be to an astronaut in some unspoken future invested with a quality of strangeness and remoteness. Like in moonlight, geysers, craters, and frost would appear to lose their actual substance, and become things of intellect alone.

Up above, at least five moons loom. What the smaller ones, Styx, Nix, Hydra, and Kerberos, might look like is uncertain. Bright lights rising and setting. But Charon is another kind of monster. Our moon spans 0.5 degrees in our sky, large enough to seem mysterious, while Charon spans 3.5 degrees of arc in Pluto's sky because although smaller than our moon, it's much closer to Pluto's surface. And unlike the Earth-moon system, Pluto and Charon are locked in one-to-one orbital resonance, meaning Charon's orbital period is exactly the same as Pluto's rotation period: 6.387 Earth-days. It hangs unmoving over one hemisphere of Pluto.

For in technical terms this is not a planet-moon relationship. It is a double-planet, or maybe double dwarf-planet, system. Charon's diameter is more than half of Pluto's, and it was massive enough at the time its progenitor (most likely) wandered into Pluto's gravitational vicinity and crashed that the center of Charon's orbit is not Pluto itself, but a gravity node just above Pluto, around which they both orbit.

To an astronaut in that unforeseen human future plodding over Pluto's twilight ices, or to the imagination at work in Maine's woods, Charon would be a giant glimmering nightmare. A huge, dimly lit shadow because it reflects so little sunlight, seven times larger in the sky than our moon, looming motionless, caliginous-seeming, waiting. It might become an emblem, even in the mind of an astronaut, of the experience of Pluto's frost-ridden surface, beetling shadows, and icy gleam. Such monstrosities of sky and cold might simply terrify an explorer in a spacesuit.

Or they might not. During that first moon landing in 1969, Buzz Aldrin famously described the lunar landscape as "magnificent desolation," an oxymoron on an experience of profound beauty in perceptually disorienting topography. The astronauts had no context for judging how large or close the lunar boulders and mountains were. The light, shadow, and perspective were strange and remote.

Pluto is much farther away and colder than the moon, much deeper on the edge of science's epiphenomenal inability to connect physical reality to human experience. What a Pluto-walking astronaut might actually experience is not clear. It is not well-mapped territory.

January

* * *

Under Pluto's surface there might be an ocean of liquid water. This hypothetical detail provoked maybe the most excited speculation ahead of New Horizons' encounter. "The subsurface ocean question is central to everything that we don't know yet about Pluto's geology," said one science writer in 2013.

The ocean question is central to everything unknown about Pluto's geology because it would solve the problem of Pluto's internal structure. If inner Pluto is an undifferentiated mix of rock and ices, that's one thing, with one set of implications for how Pluto formed and evolved to its present frozen corpse-like state. But if its interior is differentiated into a water layer and a rocky core, that's something completely different. It would mean, for one thing, that there is a significant heat source.

Water gets the attention of astronomers and astrobiologists because it is a sort of common denominator. It's central to everything we know about life on Earth. All living organisms, as far as we know, have water at the source of their biochemical processes. We came from water. Our bodies are made mostly of water. One of the clichés that justify the expenses of space exploration is that it helps us learn about ourselves. So water is an elemental link. If an ocean of water exists under Pluto, then we will have found a fragment of our own source reflected there.

The astrobiologists have all but ruled out the possibility of life in any putative Plutonic water. The necessary organic nutrients are thought to have long since leached away. But an ocean inside Pluto could be a sign of water on other large KBOs, like Eris. And they might still have their nutrients. Is there life in the solar system's borderlands?

Key to knowing whether liquid water exists inside Pluto is knowing what process might be melting ice. For there has to be a spark that raises the internal temperature. There are five possible ways planets (and dwarf planets) generate heat: they may have heat left over from the energy released during their crash-battering creation; heat released when denser materials churn downward and less dense materials churn upward; heat released in radioactive decay; heat from tidal distortions like those that the sun's and moon's gravities cause the Earth and that Jupiter's causes Europa; and heat released when liquid solidifies.

It is known that Pluto gives off heat. It's measured at a rate about 1 percent that of Neptune. The planetary scientists' prognosis is that the heat is generated either from creation, or from churning materials, or in particular, from radioactivity. If an isotope of potassium is decaying in the core, then a temperature of around 1,340 degrees Fahrenheit might be generated. Enough heat to keep an ocean (containing some ammonia acting like antifreeze) liquid.

The discovery of a heat source inside a place where hell has otherwise frozen over would constitute a wealth of information.

* * *

Pluto's name was suggested by the eleven-year-old granddaughter of an Oxford University librarian. When Venetia Burney Phair's grandfather Falconer Madan read a newspaper account of the new planet's discovery at the breakfast table in March 1930, Venetia, having some knowledge of Greek and Roman mythology as well as some schoolgirl astronomy, said, according to the accounts, "Why not call it Pluto?" Grandfather Falconer's brother had named Mars' two moons, Phobos and Deimos, and liking Venetia's idea, Falconer sent the suggestion up his line of academic contacts. It quickly reached the Lowell Observatory in Arizona, where it had already come up for discussion, and was approved. The ninth planet was Pluto.

Somehow Venetia's guess, or intuition, or epinoia, worked. Pluto is indeed distant and shadowy, qualities we conventionally ascribe to the underworld, whether taken for a real place or a fantasy. The astronomers were pleased with the word "Pluto" because they by and large view the names of astronomical bodies as entertaining whimsies. Scientific common sense presumes that ancient people, bright as they otherwise might have been, were nevertheless vastly ignorant of physical facts and so were restricted to concocting names by fantasy associations. You can find on the Internet, for example, any number of reliable scientific authorities rehearsing the idea that Mars was named for the god of war because it appears red, and red implies blood, and blood implies war.

It's doubtful, however, that the ancient astronomers were really so superficial. An ancient name, like a planet's geology, is nowhere near as simple as its color. Certainly, the ancients were no less intelligent and no less diligent than our scientists, despite our mighty telescopes and advanced mathematics. Cave paintings in southern France that were made thirty-two thousand years ago reveal, at very least, proficiencies of stroke, line, composition, and color to rival any artwork made in historical times. But those ancient geniuses did not possess and were not seeking the same kind of knowledge modern astronomers possess and seek. The astronomers of five thousand and more years ago not only noted a star or planet's color and tracked its path, they also paid attention to the particular kind of awe its luster worked on the mind's eye. They would have been as interested in a Pluto-walking astronaut's apprehensions of looming Charon as the geology beneath his feet. Probably more so.

Venetia, little as she was aware at the time, accurately expressed certain implicit details of Pluto no one else could know either. The name appears "to fit into archetypes that are already there," the astrologer Phil Poirier said in a conversation, and those archetypes involve that which is shadowy, unknown, at the edge of human control, and transitional to previously unseen realities. Was the twentieth century uncovering of horrors, Poirier asked, from the Holocaust to sexual abuse in the church to the burgeoning of mental health therapies, merely coincidental to the discovery of Pluto and its approach to perihelion in 1989? To the scientific mind the answer is resoundingly: yes, of course. On the other hand, a planet (or dwarf planet) has a presence in the mind beyond its rock and ice. Whether that presence is simply a metaphor, or something more fundamentally interconnected to experience, in either case those mental experiences are more than whimsical associations to a color. They are part of a continuous texture.

"The sky is just the same as the earth, only up above, and older," a Micmac storyteller told an anthropologist around the turn of the last century, distantly echoing a certain kind of knowledge from ancient times.

Pluto does not mean hell, exactly, although that's the conventional notion. It's easy to daydream of metaphorical Charon sculling his boat nearby, waiting for the newly dead. But when you get inside the word, you discover that Pluto is not the underworld place, but instead the ruler himself of Hades. In ancient Greece, the underworld was the place where valuable minerals were found.

"Pluto gives wealth (ploutos)," Socrates explains in Plato's *Cratylus*, "and his name means the giver of wealth, which comes out of the earth beneath. People in general appear to imagine that the term Hades is connected with the invisible, and so they are led by their fears to call the god Pluto instead. ... In spite of the mistakes which are made about the power of this deity, and the foolish fears which people have of him, such as the fear of always being with him after death, and of the soul denuded of the body going to him, my belief is that all is quite consistent, and that the office and the name of the god really correspond."

To know what's inside Pluto might be central to everything.

* * *

You need a telescope to see it. In July 2015, it's nearly thirty-two astronomical units away and reflects sunlight back at magnitude 14.1. By comparison, Uranus and Neptune, at magnitudes 5.8 and 7.8, are bright enough to be seen on dark, clear nights with binoculars, if you know how to read the star maps.

For us in the early twenty-first century Northern Hemisphere, Pluto is meandering the southern sky in Sagittarius, which is just about the overall most light-filled constellation because in that direction lies the fullness of stars at the galaxy's nucleus. On July 14, 2015, Pluto is a bit northwest of that center. It's just below Venus, which that morning is near its boiling brightest at magnitude minus 4.7. At another time of year you could, with some determination and a 4½-inch reflecting telescope, spot Pluto near the x Sagittarius stars (36 and 37 Sgr, with naked-eye magnitudes of 5.1 and 3.5). At this time of summer, though, the sun is about to rise in northern latitudes, and dawn washes out almost everything but Venus.

Even in the telescope accurately aimed from your back-yard clearing, Pluto is just a tiny speck of light among a million others. And so no matter whether you can match one crystal-cold dot in a lens to a dot on a star map, or just esti-mate its general location with your eye, its presence is largely a figment of your imagination. Your imagination can travel in a lot of directions.

The most well-worn of those in our eon is mapped by ro-bots. Voyagers I and II and Pioneers 10 and 11 years earlier cleared the elliptical vicinity of Pluto's orbit. Voyager I in 2015 is beyond the defined range of the Kuiper Belt and drilling along through interstellar space. When your mind is imagin-ing it out there like this, you can kind of keep a grip simulta-neously on the feelings of awe and trepidation that thoughts of deep space inspire and on the basic physical data, such as the figures that describe distances: Voyager I in March 2014 was more than one hundred twenty-seven astronomical units from the sun. That is a graspable number reflecting a chilling kind of isolation.

At some indefinite psychological intersection, the objec-tive fact meets the mind and gets a shape. One of the shapes is rational: it goes into a category and becomes part of a de-scription of a physical process or condition. Another of the shapes is emotional: the image of deep space inspires a feeling of cold awe, possibly fear.

Science eschews emotions. They are obstacles to rational clarity. Trepidation, awe, fear, hope, desire, pity, animosity, avarice, and so on act like smears on the telescope lens of ra-tional thought and you can get lost in them as if in pathless woods. You do not want your astronauts walking around on the moon, let alone on Pluto, frightened out of their wits.

On the other hand, part of the reason you go out to the backyard with your little reflecting telescope is to put yourself in the way of awe and trepidation. You can't, after all, gather any details that the astronomers with powerful robots haven't already got.

The Pluto of awe and trepidation is a shape in the mind that might have gripped Edgar Allan Poe's imagination. Floating up there somewhere in Sagittarius, a figure of darkness, shadows, and icy desolation. For Poe, beyond the fundamental, even adolescent-seeming horrors of his stories, was one of the Western world's most inventive proto-psychologists. He made some of the first modern maps of differentiations in the human mind, predecessors of Freud's conscious/unconscious layers and Jung's core archetypes and collective unconscious. Poe's maps of our mental geology show that the experience of Truth is processed through the Intellect, whose faculty is Reason; the experience of Duty is processed through the Moral Sense, whose faculty is Conscience; and Beauty is processed through the Soul, whose faculty is Taste; what Poe calls Passion is parallel to what he, and we, somewhat nebulously refer to as the Heart.

In Poe's view of the human mind, the objective, scientific facts of Pluto might inspire feelings that correspond, in a phrase made by an explorer, to magnificent desolation, where the intellect grasps the desolation, and the faculty of taste (in its nineteenth century sense) experiences the magnificence.

Astronomers, however, bypass these experiences as besmearing emotions, and neuroscientists discard this kind of psychology as, well, immaterial. For the only concrete evidence of mental or emotional experience is in tabulations of biochemical activity. And since the only physical evidence of the existence of mental activity is in chemicals, it is confidently assumed that the chemicals generate the activity. In other words, thoughts, emotions, and awareness itself are produced by biochemical activities. Consciousness is a by-product of, or epiphenomenal to, chemical processes.

But if chemical activity in the brain is generating the awareness of physical perceptions such as cold, starlight, salt, thunder, and sulfur, then what is happening when the objective facts meet experiences and faculties inhabiting reaches of the mind beyond the range of raw emotion? If outer physical events trigger chemical reactions that result in sensory perception, then what triggers the chemical reactions that register nonsensory, mental experiences such as beauty?

Plenty of scientists, going back at least to Whitehead and Eddington, know very well that the five senses detect nowhere near the whole range of physical reality. We perceive only a tiny fragment of all the wavelengths of light that comb the universe; no one has ever smelled a quark. Despite the best efforts of modern science to deny or ignore them, signals other than five-sensory are coming out of the cosmos, and religious adepts explored and elucidated their sources long before Poe.

The astronomer Robert Jastrow a few decades ago gave a parable of scientific knowledge's link to religious knowledge: a modern scientist after a lifetime of work finds he "has scaled the mountains of ignorance; he is about to conquer the highest peak; [then] as he pulls himself over the final rock, he is greeted by a band of theologians who have been sitting there for centuries."

Once you climb out of the foothills, names are no longer whims, and metaphors are no longer toys. About two-thirds of the way up Mount Purgatory, Dante turns to speak to his guide, Virgil—who signifies the strength of rational mind required to navigate the horrors of hell and the disciplines of purgation—only to find that the guide has peeled off and disappeared because his rational intelligence serves no purpose further up. There is another kind of knowledge at the mountaintop, and it is encountered in Beatrice.

Beyond the emotional and then beyond the rational are ranges of experience in a psychic borderland difficult to describe because it is, in a way, so distant from everyday perceptual reality. It is for the most part so far off scientific maps that neuroscience denies it even exists. Its images seem shadowy and disorienting. It can be described here no more clearly than as the silence that emerges when noise subsides, or as the realization that the image in a mirror, very clear to your physical eyes, is empty. Not the rational or figurative concept that the mirror is empty, but the realization of its emptiness.

* * *

Out of the sky and down through the firs into the tube of the little telescope flows starlight. It strikes a mirror at the bottom of the yard-long cylinder and bounces back up onto another mirror that reflects it again into a lens that funnels photons into your eye. Inside your skull the light signals transpire into a texture of the cosmos. It's a state of mind that stargazers deliberately seek.

At first, on a July pre-dawn, that texture is experienced like a door opening into fresh air. It unfolds into the feeling, maybe, that what's above is the same as what's below, only older. Much older. The overarching feeling, whatever else goes with it, is that the sky is unspeakably beautiful.

The light you're seeking this July morning is hidden in the borderlands of the solar system among a millionfold fullness of vastly more distant lights. Sun rays travel three billion miles, strike Pluto's frost surface and bounce back Earth-ward to a 4½-inch mirror pointed toward the galactic center. A tiny point of reflected light pierces your eye.

January

Streaking past Pluto this morning is a robot so tiny in comparison to a moon that its reflection can't be detected by telescope. But built into it is an artificial life of radioactivity and electricity configured to collect signals from the dwarf-planet system and send them on beams of light back to Earth with arcane mathematical coherency. It's the same coherency that emanates us-ward in starlight, arranged so it corresponds to the wavelengths we detected first by eye and then deduced are signs of further wavelengths that transcend the eyeball. It's as if we sent a replica of ourselves out there. The robot picks up data and reflects them back toward Earth, mirroring the processes of perception and communication. It carries the same radioactive spark that probably liquefies water inside Pluto, where under slightly different circumstances life might blossom.

Nothing the robot sends back can be detected by the eye among the pre-dawn firs, even with a telescope. But something else is bouncing among the tiny Earth-bound telescope's mirrors that the robot can't detect. Signals of vastness, loneliness, and power reflected in as if from a sentinel on the edge of everything.

Pluto, though, to us is lifeless. A remnant of ancient warfares among planetary gravities. Whatever the robot finds will be frozen, dead matter. A corpse, in a way. Its very epinoian name reflects the ruler of the place we all descend to: dark oblivion, or the energy that coheres some eonic afterworld from whose bourn no traveler returns. The robot is configured to find dead matter only, and being no more than a reflection of a human being, can't imagine anything else. The human scientists are trained to dissect, analyze and describe the findings. They hope to find inside the body an interior spark that—hope upon hope—will help us learn about ourselves. Inside Pluto is a wealth of information.

"Let us not underrate the value of a fact," Thoreau warned; "it will one day flower in a truth." He meant that the physical world is a reflection of nonsensory cosmic textures; and he thought of those textures as the moral world, which itself is a kind of foothills or borderland to ranges further up the mountain where some monks and enlightened scientists are taking a rest. They see that matter, a momentary configuration of energy, is essentially empty. If they followed the conventions of science they would become mired in dissecting details and find, to paraphrase an ancient mystic, nothing more than a corpse. The facts, isolated from the cosmic texture, are empty desolation. But when the faculty that detects beauty recognizes that emptiness, then silence blossoms into a clarifying voice, silently shining to silent moons.

New Horizons, unbeknownst to itself, has gone to the borderlands on a quest for gnosis.

Winter Midnight

"… all night long I feed the stove
birch logs like thigh bones, stir up hot embers,
the constellations
step through the window
and sit in the rocking chairs,
our only visitors."
        —Mark Rutter, "Trick Staircases"

Midnight. January. The back deck.
It's just me, the snow, the cold, the pointed firs, and the black sky where numberless crystal points of light are burning. Sirius, Betelgeuse, and to the southwest reddish Aldebaran, in the constellation Taurus.

Every year this all seems deeper and larger than last and, almost unbelievably, more expansive. Most of it you can't see until you square the facts with your mind. That can be a bewildersome undertaking this deep in winter.

Maybe you know that at just about midnight on New Year's Eve 2018, the New Horizons spacecraft flew its obscure, lonely route past the space rock called Ultima Thule, the most distant object ever seen up close by human robot. Object 2014 MU69 is about four billion miles away in the Kuiper Belt, the wild, weird clime that lies beyond Pluto, in the direction of Sagittarius, which this time of year is behind the sun.

Deeper. A little more than a year earlier, from out of space and time came wheeling into the solar system the first object ever seen whose trajectory indicates it's from interstellar space. The astronomers tracked it carefully. It's highly reflective, appears to be oblong in shape, and is tumbling. They don't know what it is. It has the wrong shape and path to be an asteroid. They're pretty sure it's not a comet because it did not develop a tail of outgassing material when it slingshotted around the sun. And most strangely, when it should have been slowing down as it left the sun's vicinity, it accelerated. In a report filled with detailed mathematical analysis, some astrophysicists concluded there's a strong possibility the object is "a lightsail, floating in interstellar space as a debris from an advanced technological equipment." In other words, professional astronomers think it could be a piece of an alien spacecraft. The astronomers named the object 'Oumuamua, a Hawaiian word meaning "scout."

Larger. In the 13.7 billion years we know of that the universe has existed, around two trillion galaxies have arisen. We live in one of them, and you can easily see another one, M31 in the constellation Andromeda. It looks like a faint smudge in a clear, dark January sky. Recent studies indicate all these galaxies have contained about a trillion trillion stars, and taken all together, these stars have generated 4×10^{84} photons of light. Or to spell it out for arithmetic, 4,000,000,000,000,000,000,000,000,000,000,000,000, 000,000,000,000,000,000,000,000,000,000,000,000, 000,000 photons.

The astrophysicists say nothing about your actual experience of a fact like this, about what your mind can or cannot do with an enumeration of particles of light. Some of the problem is that in an expression like "the universe began 13.7 billion years ago," there is really no such thing as "ago." At a certain point (which is much closer to your backyard than you might think), time and space congeal, and years—as in "light-years"—describe not just a period of time, but a distance in space. The universe is filled with light, from start to finish, or really, just simply forever. All the stars above my firs are just the ones near and bright enough to see by human eye in this one galaxy lying in its inconspicuous nick of time.

All this has been on people's minds much, much longer than has historically been believed. Researchers building on other studies in archaeology and astronomy recently verified that the distant antecedents of people who seventeen thousand years ago made astonishing cave paintings in southern France were well aware of a twenty-six thousand-year cycle in the stars which we call the precession of the equinoxes. Their lore and learning contained factual information—in forms much different from scientific reports—describing what's known now as coherent catastrophism: periodic collisions of cometary debris with the Earth that cause worldwide climate and ecological changes. Ten thousand, seventeen thousand, forty thousand years ago, people knew about these disasters, and knew they were associated with the sky in the direction of the constellation Taurus. Even though they did not know about meteor streams left over from shattered comets, they had ways of remembering and describing them.

Aldebaran and the other stars of Taurus, just west of Orion, are still up there on January midnights. They're the same stars my ancient ancestors saw in northern Europe. It all expands and congeals, somehow, in a single moment on the back deck. Out of space—out of time.

Ursa Major (Preview)

Here in Maine, where we are just about halfway between the equator and the North Pole, the Big Dipper dominates the northern sky all year round.

If you watch it all night, you'll see that in each twenty-four-hour period the whole constellation revolves completely around Polaris (the North Star) in a giant circle. If you keep track of it even casually for a year, you'll see that it moves through the same circle over the course of twelve months; the nightly circle is embedded in the yearly circle.

We call it the Big Dipper because its seven brightest stars visually resemble a ladle. A thousand years ago, Britons called it the Wain because they thought it resembled a wagon. At least, so goes the astro-etymology. The scientific name, held over from Latin, is Ursa Major, which means Great Bear. The seven stars of the ladle's cup and handle are the bear's body and tail, and about fourteen other stars outline the head, neck and legs.

The Greeks at some point called this constellation Callisto. Artemis, the goddess of the hunt, changed Callisto into a bear for betraying her chastity, and Zeus rescued her by placing her in the sky. Callisto's child was named Arkas, giving rise to the Greek word αρκτοσ (arktos), meaning bear and becoming associated, at some point, with Polaris at the tip of the handle of the nearby Little Dipper, or Ursa Minor. Our word "arctic" — in the sense of north — comes from this root.

In the Middle Ages, Arab astronomers also associated Ursa Major with a bear, calling it Al Dubb al Akbar, the

Greater Bear. A vestige of these words remains in the name of the star at the top right of the dipper's cup—Dubhe. In fact the official Western names of many stars are taken directly from Arabic names—Dubhe, Merak, Mizar, Alcor, Alkaid, and even Alioth (through a series of mistransliterations), to name just the bright stars in the Big Dipper.

Not only the Greeks and Arabs saw bears in the northern sky. In Britain the wagon specifically belonged to Arthur, who was the king expected to return to power. It happens that the Welsh word for bear is *arth*, pretty clearly a root of Arthur.

The Arthurian myth in its association with the great wheeling constellation of Ursa Major evokes a sense of cosmic power and eternal recurrence. A great bear returns year in and year out to its appointed stations, revolving eternally around Polaris, which from our vantage point here on Earth is the still center of the sky.

It turns out a number of Native American cultures also call the bear a bear. The Zuni Indians of the Southwest refer, in describing winter, to the sleeping bear, and in describing spring, to the bear's awakening. Some Native Americans relate that a cosmic hunt takes place in the sky each year, involving a large game animal, like a caribou or elk, and often, a bear. The Coeur d'Alene and Thompson Indians of the Northwest describe the stars of Ursa Major as a grizzly bear and hunters, as do the Cherokee. In the Northeast woodlands, Ojibwa, Menominee, Fox, Illinois, Iroquois, Delaware, and Wabanaki cultures all associate the stars of Ursa Major with a bear that in most cases is being hunted.

In winter, goes a Micmac version, the bear (represented by the four bright stars of the dipper's cup, upside down) is on her back in death. Late in spring she begins to stir, as about midnight the stars in the cup are beginning to tip down. The stars in the handle are hungry hunters who decide to chase her. All summer they pursue her, their track bending down-ward. Some of them lose the trail and disappear below the horizon. By autumn, the remaining hunters corner the bear and kill her. Her blood spatters the autumn leaves. The bear dies, sleeping her death sleep through the winter. In spring, the cycle begins again.

You can look up there any clear winter night and see that giant bear circling the axis of the cosmos, settling into her win-ter sleep at this time of year. When you're dead and inhabiting the stars yourself, those thermonuclear bears will still be hanging there over northern trees and trees to come, undis-turbed and wheeling. Returning every spring to life on Earth.

The Great Bear in Maine

In 1977 I climbed Knocknarea, a big round hill outside the town of Sligo in Ireland, to see Queen Maeve's tomb. For the better part of the morning I followed stone walls, tramped across sheep fields, and kept thinking the top of the hill wasn't getting any closer. It turned out to be much farther away than it looked from a distance, but the weather was mainly pleasant, with bright fall sunshine and occasionally a huge black cloud that dropped gouts of rain for five or ten minutes. It was too warm in the sun and too cold in the rain, and so I alternately took off and put on my red-and-black wool New England hunting jacket.

At the edge of the hilltop, a horizontal gully of sand pushed under a lip of grass like a natural earthwork protecting the summit. As I pulled myself over it a gust of wind blasted my face. I clambered onto the top and saw an immense pile of loaf-size rocks, more than thirty feet high and at least two hundred feet in diameter—though I came to know only the mound's height, and not its circumference, from experience.

This was Queen Maeve's tomb. It has been there for five thousand years or more. Archaeologists assume that a passage tomb, like others in the area, lies underneath it, but no one knows for sure. In fact no one knows what the pile—or cairn, as it's called—was ever for at all. The archaeologists and other scholars doubt the local explanation that Queen Maeve, a warlike figure of Irish legend, is buried there. The stony ruts and depressions of the Irish landscape are said to be inhabited by spirit-people and faeries, and this pile of rocks, huge and startling, is naturally the center of a lot of stories, tales, and other outright lies that make up Irish mythology. Irish myth is peculiarly different from the more intricate mythologies of the rest of Europe. It seems to originate in the rocks and soil and peat. European and Middle Eastern myths more often derive from the sky and stars, or from the sea in its associations with the sky. There is something familiar to me, a New Englander, about Irish geography and its inhabitants.

As I stood for a moment on the hilltop a big black cloud angled in from behind, and so I set my small canvas shoulder bag down on the grass and pulled my jacket on. I decided to wait out the rain, which was only a minute or two away, by finding a flat-sided rock on the edge of the pile and sitting down to eat my standard traveling lunch of bread and peanut butter. The wind was wicked, as we say in Maine. I turned my collar up and fumbled with the wrapping around the bread, and had to search through notebooks and books and pencils to find the jackknife I used to spread the peanut butter.

Then it began not to rain but to hail. The wind gusted and swirled, and I literally felt like it was getting ready to carry me off. Big frozen chunks—some were snowflakes and others were ice crystals—plastered me in the face, fell in my carton of milk, and actually began to accumulate in the folds of the canvas bag and around the rocks at my feet. I closed the flap to protect the books and huddled to eat and wait for the snow to pass.

I didn't eat much. The wind seemed to circle the mound and blew down inside my jacket. In fact it was so strong and irregular that it arrested my attention, and I was no longer aware of the sky but only the blowing and the immediate earth around me, the rocks and ragged grass. I had to struggle to focus my eyes on the shape of the mound and the lie of the rocks. I wanted to look over toward Sligo and the bay in the distance, and maybe climb up to the top of the tomb, but instead all I could feel was the wind. I looked not at the rocks or the expansive Irish countryside around me, but directly at the air, as if it was an object. It was as if I could no longer see the sky. My vision fuzzed and my eyes were watering with a kind of wind-blindness—I was aware of, but could not focus on, the ragged sky. The wind created the weird sense of being in deep, tangled woods, even though I was on a high, open hill. The wind became everything, and its single point of reference was the pile of rocks around and above me. I began to have the distinct, almost hypnotic impression that I was being warned off the hill. This wind was abnormal. I became spooked.

When the snow passed I left the bag on the ground and decided at least to walk the edge of the mound. But as I took four or five steps, the wind picked up. In fact the farther around the pile I walked, the harder the wind blew. So I walked back to my original place, which seemed safer, and the wind lessened there as if I was on the leeward side of the tomb. I picked up the canvas bag and tried to walk around in the opposite direction, but this time the wind gusted hard enough to wing my hair back and forced me to lean forward to keep my balance. It roared against the rocks, I remember whistling noises. I took several steps over the stones scattered on the ground at the base, but finally realized I was virtually closing my eyes, the way you would in a wind tunnel.

At this point I began to wonder if the spirits of the tomb were at work. I had the impression that I was surrounded by the wind and if I insisted on walking the entire circumference of the mound, I would be in danger. At least, it seemed safer to follow my superstitions than it did to maintain a modern objectivity that might in the end get me blown straight off the edge into the sheep pastures.

I tried one more time to go around the other way, but the wind whirled full force, rushing directly at me. I turned back and walked away from the huge tomb, and when I stepped over the lip and onto the sand on the hillside, the wind ceased, utterly. Suddenly there was no wind to lean into. The stillness staggered me, almost like a whack with a club. I was in placid, sunlit calm, and my head wheeled with the change. Feeling I might keel over, I sat down on the grass until I could see again, looked across the pastures down to the blue-gray road in the distance and toward the town and its spires beyond. I silently repeated a verse from Yeats I had been memorizing in bed at night:

> *The host is riding from Knocknarea*
> *And over the grave of Clooth-na-bare;*
> *Caoilte tossing his burning hair,*
> *And Niamh calling Away, come away*

On the walk back I came across a dolmen and wondered what it was ever built for. Earth spirits, or psychologies, are enclosed and secretive, and work their way downward. Sky psychologies are expansive and upward. At the summit of Knocknarea the impulse to look up clashed with a weird necessity to focus downward, tomblike, and the mixture kicked up a whirlwind. Something like that. It made a working story, or metaphor, in my mind for an extremely strange and somewhat frightening couple of hours.

* * *

All this about Queen Maeve's tomb must seem pointless,

given the title of this essay. I want, after all, to unfold the night sky of Maine, not Europe. Although I have seen and heard, or thought I heard, some strange things in the Maine woods, I never had the same feeling I had on Knocknarea, until one particular winter day about fifteen years later.

I've wondered about the fact, apparently, that the Native Americans of the Northeastern woodlands did not develop sky mythologies as extensive as other cultures'. If you walk around in the woods often enough, you can tell at least a few superficial reasons why. The Maine woods are close and dense, and cold eight months out of twelve. Pine, spruce, cedar, fir, and hemlock sixty and seventy feet high—even higher a millennium ago—obscure the sky year round. In the prehistoric boreal forest there were few panoramic views of the stars. You had to deliberately climb a granite mountain and come out on a ledge or a big bald spot, or camp directly on the shore of a large body of water to gain the big picture. Ages ago it was more usual to lie on your back and gaze at twinklings refracted through fir boughs than to study a clear, open field of stars as you might in Greece or Persia.

I say this in full knowledge that some constellations shine and move with unmistakable clarity and power everywhere. Orion with his jewel belt. And in the Northern Hemisphere, the Big Dipper milling eternally, night after night, year in and year out around the North Star. Among all the stars and planets, the Big Dipper gives the clearest nightly picture of where you are, how far along the journey from winter to summer you are. And not to be too clinical, but before Europeans arrived, the stars in North America were not exactly stars—not isolated, hopelessly distant thermonuclear fires, at any rate. They were powers understood, like time, to be an intimate element of everyday life. The Big Dipper was a cosmic clock, but—and this is hard to get, even I don't have it as clear as I'd like, with my glowing digital clock and a calendar filled with numbered squares beside me—a clock but not a timepiece: it was more like a revelation of moments. Revelations are not understood the same way time is understood. Revelation is experience, not fact.

In ancient North America, time could not be told by a watch. But it could be told by a story, which reveals not only a moment in the movement of the seasons, but also the experience of that moment—the feel and significance of it that goes well beyond the accuracies of azimuth and altitude.

Myths are, we might say, endemic to the world, and the stories about the Big Dipper show this. We call it the Big Dipper because its seven brightest stars visually resemble a ladle. A thousand years ago, Britons called it the Wain because they thought it resembled a wagon—at least, so goes the theory. But poking a little past the merely visual, the Big Dipper's scientific name, held over from Latin, is Ursa Major, which means the Great Bear. The constellation covers not only the seven stars of the cup and handle—which make up the bear's body and tail—but about fourteen other stars outlining the head, neck, and legs as well.

The Greeks at some point called this constellation Callisto. Artemis, the goddess of the hunt, had changed her into a bear for not remaining a virgin, and Zeus had rescued her by placing her in the sky. Callisto's child was named Arkas, giving rise to the Greek word αρκτοσ, meaning bear and becoming associated, at some point, with Polaris, the North Star at the tip of the handle of our Little Dipper, Ursa Minor. Our word "arctic" derives from this root, and an echo of the ancient association of bear and pole—the central axis around which everything turns—sounds.

It's an interesting story, from a certain perspective. There's a deeply felt response to the complex relations between

Callisto, Artemis, Arkas, and Zeus which has been preserved for millennia. Maybe the constellations are a chance mosaic or a simple mnemonic for a deeply felt story. The stars have many ways of inspiring awe.

Arab astronomers also associated Ursa Major with a bear, calling it Al Dubb al Akbar, the Greater Bear. A vestige of these words remains in the name of the star at the top right of the dipper's cup—Dubhe. Arkas is also heard in "Akbar," a clear indication that there was some link between Arab and Greek mythologies. This link does not seem far-fetched to us here in the historical world. Greece had extensive philosophical, religious, and scientific intercourse with the Middle East, and in some ways has always been more Eastern than Western in its culture. In fact the official Western names of many stars are taken directly from Arab sources—Dubhe, Merak, Mizar, Alcor, Alkaid, and probably Alioth, to name just the bright stars in Ursa Major.

Not only the Greeks and Arabs saw bears in the northern sky. In Britain the wagon was known specifically as Arthur's, the king expected to return in his power. (Later, during the Renaissance, the constellation was called Charles's Wain to keep the association of power with the king.) Now, it happens that the Welsh word for bear is *arth*, pretty clearly a root of Arthur. The Arthurian myth in its association with the great wheeling constellation of Ursa Major evokes a sense of cosmic power and eternal recurrence. A great bear perpetually returns, year in and year out, to its appointed stations, revolving eternally around the North Pole, which is clearly the still center of everything,

Believing that the bear myth somehow traveled overland from the Middle East to the British Isles does not require a leap of faith either, really. Northern European cultures contacted Mediterranean cultures extensively in ancient times, long before Hadrian built his wall across Northumberland in the early centuries A.D. We begin to be talking about stories that must be very old, if it's the case that Celtic people shared a story, or a version of a story, with Middle Eastern cultures.

Did ancient storytellers pass the fabricated meaning of the star-mosaic from culture to culture so that it spread northward (or south)? We love causes and effects that happen neatly and linearly in time. But something feels not quite right about it. Stories reveal experience, really, rather than simply transmit "information." The identity of star myths and names in different cultures feels, inside its linguistic form, more complex than the repetition of a story. The reason I say this is that not only Arab, Greek, and Celtic people associate the Big Dipper with a bear: so have the Chinese. It begins to stretch the credibility of cause-and-effect rationality to think that the Chinese bear's story is related to Ursa Major, Callisto, Al Dubb al Akbar, or Arthur. And yet there it is. Also in Japan.

And also, it turns out, in a number of Native North American cultures. In saying this, we are speaking of cultures presumed to have had no significant contact with either European or Asian people before about the fifteenth century. The Zuni Indians of the Southwest refer, in describing winter, to the sleeping bear, and to its awakening in spring. The Coeur d'Alene and Thompson Indians of the Northwestern Plateau area describe Ursa Major as a grizzly bear and hunters, as do the Cherokee, who occupied the vicinity of North Carolina, thousands of miles away, when Europeans arrived.

Certainly, not every culture views Ursa Major as a bear, but many do, and others identify it with a large game animal like a caribou or elk. Their stories also relate that a hunt takes place in the sky each year. In the Northeast woodlands, Ojibwa, Menominee, Fox, Illinois, Iroquois, Delaware, and Abenaki cultures all associate the stars of Ursa Major with a bear that is usually being hunted, as Artemis herself was a hunter.

In winter, goes a Micmac version, the bear (represented by the four bright stars of the dipper's cup, upside down) is on her back in death. Late in spring she begins to stir, as about midnight the stars in the cup are beginning to tip down. The stars in the handle are hungry hunters who decide to chase her. All summer they pursue her, their track bending downward, and some lose the trail and disappear below the horizon. By fall, the remaining hunters corner the bear and kill her, the blood spattering the autumn leaves. The bear dies, sleeping her death sleep through the winter. In spring, the cycle begins again.

The cultures that associate Ursa Major with a bear and a hunt are great distances, all kinds of distances, from each other, but some visual features of the Great Bear are of course the same for whoever looks up there. At the latitudes of Maine, it's the one major constellation visible all year round. If you watch it all night, you'll see that in each twenty-four-hour period it sweeps completely around Polaris in a giant circle covering the whole northern sky. If you keep track of it even casually for a year, you'll see that it moves through the same circle over the course of twelve months; the nightly circle is embedded in the yearly circle. The Northeastern cultures described this motion in their hunting story, centuries— at least—before white settlers arrived.

Does that mosaic of stars look so much like a bear that people separated by thousands of miles and years inevitably visualize a bear? Do the stars' motions so clearly resemble a hunt that people totally separate from each other recognize it alike? Or, asking the same thing in a more scientific way, as if cause and effect were the only possible relation, astronomer David Wilson wanted to know: "Did a culture of Mediterranean peoples reach the Americas long before Columbus? and did they teach the 'bear' to American shamans?"

The truth is often simple, but never superficial. Is there a link between the Dawnland people of eastern Maine and Canada, and medieval Arab astronomers? Is it all, as a modern explanation might imply, a vast coincidence based on the resemblance of a chance pattern of stars to a bear, or their annual motion to a hunt?

No one will ever be able to say with scientific certainty. The stories originate, that's about as much as you can accurately, rationally suggest. Where they originate may have less to do with geographic locations than with the awe the stars inspire—probably no more than a glint of the reality that reaches us when ancient starlight strikes the mirrors in our eyes.

* * *

Living in the Earth must be markedly different from living in the stars. The Greeks, Arabs, and most European cultures have studied the stars and planets and their motions in painstaking detail for millennia. Even chthonic myths turn on contrasts in the seasons, which are measured by the positions of the sun and stars as the year waxes and wanes. Western mythologies, in general, explain and depict our separation from the stars, heaven, empyrean, the center of the universe. Dante's journey from Firenze was long and arduous, descending through Hell to the other side of the Earth, up Mount Purgatory, and then ascending through the planets, and beyond them, to the ultimate destination. Arab mystics like Ibn Arabi and Ruzbihan Baqli made spectacular journeys to transcendent places near Muhammad and Allah. Distant spiritual destinations of the human soul, perhaps similar to the figure in India of the seven rishis shining in the Ursa Major constellation.

In this way of seeing the world, from places with panoramic views of the night sky, the stars seem to reflect a crystalline purity associated with the divine, and not with the impoverished, often squalid physical world. Even though the Earth itself is in most religious views a divine creation, its poverty signals its remoteness from the timeless, spaceless place that God, Jesus, Muhammad, the rishis inhabit. The unity of the cosmos is visible through special spiritual eyes that are contacted in dreamlike showers by the faraway stars.

Life is difficult everywhere. Here in northeastern North America, winter has been long and brutal since people arrived about ten thousand years ago (by scientific reckoning). Survival for millennia depended on intimate knowledge of the surroundings. More than knowledge, actually. Survival depended on one's identity with the surroundings, one's movement in and through the woods and seasons. Shamans beat drums and tunneled to spirit places. The stars were markers.

To Abenakis, the "People of the Dawn," moving through the seasonal rounds up the Kennebec and Penobscot rivers in winter and down to the seashore in summer, the woods were their element. Everything, like themselves, was alive. The forces of nature were among the pines and in the rivers like a web. The Algonquian word for this power, which expresses itself in individual beings—in humans, animals, lakes, rocks, activities, everything—is *manitou*. Iroquois people called it *orenda*, and Siouan people, *wakanda*. The closest translation we have of it is a special meaning of the word "spirit," but the two meanings are only roughly similar and led European settlers to mistakenly call native "spirits" gods or devils. *Manitou* inheres in particular objects and does not mean "God"; spirit refers to a generalized source, or encompassing transcendental reality, and our word soul is in a strict sense unrelated to *manitou* or *orenda*.

In the Maine woods, deep powers were invested everywhere, and all the inhabitants of the forest were part of it as in a great web. Frank Speck explains that Penobscot families were connected to particular places and animals. One family, for example, identified itself with the beaver, others with the eel, squirrel, raccoon, sturgeon, or whale. Thoreau's Penobscot guide Joe Polis was a member of a bear family. "All earthly animals are the descendants of the ancestor animal in the sky," goes the paraphrase of the Micmac storyteller.

Such a sentence might almost be understood in terms of the Platonic forms. It rings distantly of the Middle Eastern and Western sense that the material world is related to an unseen transcendent world, spoken of in Neoplatonic philosophy as "soul" or "love," which binds the physical world.

Well, something like that. From Plato onward, and probably long before him, there was a great illusory divide between divine and material realities despite a deep intuition of their essential unity.

In the Native American view, Earth and sky are the same. They're one place, a single cosmos, virtually identical. A human being, a clan of human beings, is indeed a bear. In the words of the Micmac story, "The sky is just the same as the earth, only up above, and older." The *m'teoulin*, or shaman, of the north woods worked familiar spirits called in Penobscot the *bao'higan* through the essential relation of everything, not only sky to Earth, but also creature to creature. As in Penobscot families, the *m'teoulin* identified with the animal and its power. None of this, I should say, involves either whimsical fantasy or superficial need; it expresses a sense of direct links with surroundings and cosmic sources: "In all things as it was and is in the sky, so it is on earth," says the Micmac story.

But where Dante and Ibn Arabi were schooled to look for pure power in a separate source, normally skyward, to catch glimpses of the unity of the cosmos, the Penobscot making his way alone through the pine cathedral inhabited power—the same power?—in the forest itself. Not star paths in the sky, but bear tracks in snow. Swollen river. Hemlock stems broken in unusual ways. The peculiar bends, twists, and deadwaters in a stream wending its way down a spruce and birch encrusted hillside. The texture of ice in a puddle. Drops of blood frozen in a footprint. A white feather or scapula. The bear who, as you can see for yourself, has traveled past your campsite in the night is the descendant—the same bear—as the one circling the pole star year in and year out. The small knowledge revealed in a swatch of pine needles disturbed in just a certain way reflects the sky itself.

People too, individuals, can be as substantial and close as the woods. The Penobscots and Passamaquoddys do not have the immense oral tradition of other cultures, even though their stories of Glooskap and the eternal twinkling bear hunt are as entertaining and instructive as anything. Places are like recordings. What is put into them, they speak.

The Maine woods are not a talkative place. Things are subdued here. Winter is long, harsh, and muffling. Summer is short, sometimes hot and stifling. There is not much to say during a walk through the trees behind the house. Too much talking quickly begins to seem like an imposition. An imposition on what, is not exactly clear, and the age of objectivity has brought to the woods tides of hikers, bikers, and camping tourists who make as much noise as they please, and leave trash. In Thoreau's time, Joe Polis quietly led him anywhere in the forest, but went silent, completely tongue-tied when Thoreau asked him to describe the route home. It is not a matter of words. The Penobscots were generally reluctant to climb to the top of Mount Katahdin, the highest point in Maine and the first in North America to catch the morning sun, for fear Pomola would become angry and kill them. What was visible to them is not visible to tourists. The human voice influences its surroundings: some Northeastern cultures allowed star stories to be told only during winter so as not to upset invisible balances. The woods harbor both seen and unseen lives.

We of European and scientific descent don't or can't see much of that life. We think the things of the spirit are invisible and therefore distant, and in extreme cases, nonexistent. To us, a bear is an organism hidden in the forest, separate from us. Its tracks or scat are signs of its existence, little else. Less extremely, we think things of the spirit might be glimpsed symbolically through distant pinpricks of light in the sky, and we have an impulse to look upward. In this province of Western cosmology, if you want to see a bear and understand its true power, you can't see its signs in the pine needles here on the remote and squalid Earth. You look for Swedenborgian, Hermetic signs in the direction of Polaris, the stillness at the center of all motion, quintillions of miles away, and try to puzzle out how the stars whirling endlessly around that still center ever came to be seen as a bear.

Really the largest and most terrifying bears you will ever see circle Polaris night after night, century upon century, dizzying and, if you look long enough, kaleidoscopic. Relentless and inevitable circling, never dipping below the horizon. When you are dead and inhabiting the stars, those thermonuclear bears will hang there over the trees, undisturbed and wheeling.

Before science changed our view from bear to dipper, these woods were haunted by powers and spirits that muffled speech, whispered and crackled endlessly, and closed hollows and gullies with tunnel-like shadows. Cotton Mather in the seventeenth century called Maine the devil's country for reasons more tangible than native superstition. The Abenaki, long before the colonists, feared the Iroquois who appeared like wraiths from the trees to kidnap men for torture rituals and women as replacements for lost relatives. Scavenging Mohawks emerged like gusts of wind to sweep away your son or uncle, or whole family, and disappear. There were other things in these woods.

In the sixteenth and seventeenth centuries, Europeans emerged from the sea and changed everything forever. They brought technology, ritual religion, disease, and a sense of order unlike anything the natives ever imagined. The vast majority of Indians passed away. The Penobscots, Passamaquoddys, Maliseets, and Micmacs, with astonishing toughness, survived into the twenty-first century, badly maimed but alive. The last fluent Penobscot speaker of her language, Madeline Shay, died near the end of the twentieth century.

The woods no longer belong to the Indians. Federal law made that clear in 1980 when the Maine tribes received about eighty million dollars—just a million or so less than an annual tax reduction given to a local shipbuilder in the 1990s—in compensation for a real estate deal reneged on by the government in 1790. The Native Americans now live in a land very neatly divided by deeds that objectify and atomize places that in earlier centuries were identified with the powers that connected them in the shape of beaver, fish, and bear.

Many Penobscots have been Catholic for generations. The Christian God, as we well know from Dante and the Apocalypse, resides in the sky, not the woods. Earth spirits are demons. Sky spirits are angels. Whatever strange cracking or howling you hear in the woods, whatever wind or unusual rushing sounds bounce off your left shoulder, are all manifestations of evil or, better yet, a perceptual illusion or disturbance best explained by psychology. Spirits are not objectively there, as scientific logic plainly shows.

Those of us with interest in the woods that is not merely appetitive—the lust to kill deer or moose, or the impulse to pitch a tent and blast music and drink beer—pay attention as naturalists, which is to say, as amateur scientists. We are all the descendants of Thoreau, poking around in piles of pebbly scat, marveling over a peat-encrusted bone or a bright blue feather, making a cup of tea (the first and last) from arbor vitae leaves, learning to distinguish black spruce from white. We take dendrology courses, go on nature walks led by park rangers, or tramp around behind the house on freezing December afternoons with no known destination or purpose.

We view a fallen, decaying hemlock trunk with the objective eye of science and then catalog it, not for analysis, but for future emotional reference. *What do I love about the woods?* you might ask yourself a day or two later, and reply: *I love that hemlock blocking my way, with dead branches poking out of it, rotting into the forest floor, and the smell of snow and nearby living spruce and fir, and the irritated chatter of a chipmunk,* all together in a hemlock vision.

Sometimes something seeded in a rustle or a shadow in the ferns causes a shiver. You wonder what prompts this response, and more, you wonder what a Penobscot walking this same route and hearing this same sound a thousand years ago would make of this shiver. Something completely different, no doubt. There are spirits in these woods, and they feel wild when you come up close against them, close and speechless and unfriendly.

But as naturalists, we stick to nature, which is to say, its objects: maples, bear tracks, lichens, snowy rocks, dead leaves, wind in pines.

Stars.

The great celestial bodies originate from the same physical source as the blue jays—we are all literally made of stardust, I heard a learned astronomer say the other night. Maybe the sense that bears circle Polaris derives from this idea, that everything in the universe has equal origins and is related. Bears live in the sky as well as the woods. As naturalists, though, we think of the world more categorically, less figuratively than this. Everything is separate in space and time and can be classified, even though it is all the progeny of the Big Bang and subsequent supernovae. In my own peculiar case, having grown up near the Maine woods in the second half of the twentieth century and being neither a hunter nor a partier, thinking about the stars is the same as thinking about the woods. And thinking of the stars and woods is (for better or worse) identical to thinking about what other people have said about them, listening for what has been said in them. What scientists have said, and naturalists. And poets and philosophers. Penobscots on the trail. And mystics, for that matter. I am trying to make some sense out of the natural world and my place in it, and my cultural bearing tells me the stars equal the woods.

Imagine inhabiting the Maine woods and thinking the eyes that look up toward the stars can be the same eyes that look down toward the bear tracks in snow, and give both the same significance. It's like saying the Penobscots could have mythologized the constellations in the same detail as the ancient Greeks, or that the Greeks could have identified their families with animal spirits as the ancient Penobscots did. It didn't happen in these ways because the bald Mediterranean speaks different voices than the boreal forest. What you see and how you explain it is inextricably complementary with where you live. Maybe it's identical. I'm telling this story in the dead of winter, as you'll see.

We live in a confusing and synthetic age. We learn that everything is exactly what the logic of our perceptions tells us it is: an agglomeration of discreet material objects. We also learn that everything is the same: a single unending collaboration and disintegration of mindless atoms. All our experiences, outside our bodies and inside our minds, are the products of atoms becoming molecules, molecules becoming chemicals, and chemicals producing reactions such as life and consciousness. The sounds you hear in the woods, and the feeling of fear and presence, are the sum total of chemicals combusting in the brain. There are no spirits or gods. Only animals and brain-hallucinations.

Similarly for the stars. The stories told by ancient cultures about the constellations amount to "mythology," and the average scientist regards them as either quaint fantasies, or as expressions of ancient psychology, or as merely unimportant because not concrete. In the modern age the word "myth" is commonly used as a synonym for "fallacy." The truth about the stars, that is, is that they are gigantic balls of gas light-years distant undergoing virtually perpetual nuclear fission and radiating immensities of energy. Astrophysicists explain stars by explaining atoms. The largest beings in the universe are understood through descriptions of the smallest. The scientific truth of all reality is that it comprises atomic and sub-atomic particles whirling stochastically apart and together. The meaning of a star is that it is a complex arrangement of atoms.

The same is true of life, and so in some ways biology is more perplexing than astronomy. A star, at least, is not thought to be living, therefore is essentially dead, and so its constitution by nonliving, essentially dead atoms is at least consistent. But how do atoms—which are merely a chaos of activity and therefore essentially nothing—give rise to living, thinking beings who are obviously something? Scientific logic indicates it is more consistent for us to think that life, too, is nothing. Everything is nothing. And stories about constellations are in particular nothing because they are concoctions of unusually complex chemical combinations in the human brain, not natural objects of inquiry in any appropriately objective sense. This way of understanding the nature of reality releases us from all responsibility of any kind toward anything, even our own survival. Since everything is nothing, it doesn't matter to the cosmos or to any individual what stories human beings tell, or whether they tell any stories at all, or even whether human beings go on for another day.

All this is very confusing because all or most human beings have a deep intuition that they are, indeed, something, and that their surroundings are also intimately something. Nature, as anyone can plainly see by kicking a boulder, is something, however transient it is. Stars are something, and you can tell they are something because when you look up at them on a moonless night, even for a few seconds drinking up the light spattered in spectacular arrays over the blackness, you experience awe, beauty, sometimes fear, and sometimes more complex emotions. Stars affect human beings profoundly. Those effects are as real inside the human psyche as a hammer smashing a thumb or as bear tracks frozen in snow.

What is a star? What is a constellation? What is a bear track?

Science's replies to these questions are the helpful beginnings of answers. But as Thoreau pointed out, a fact is meaningless until it blossoms into a truth. If we take this point seriously, then we return to the ranges of Socrates, Jesus, and Muhammad, who are the main teachers of Western culture on problems of truth, justice, moral disposition, and invisible spirit.

Or closer to physical home, return to the ranges of the Penobscots, Passamaquoddys, Maliseet, and Micmacs—somewhere along the banks of the Kennebec or Penobscot rivers, or in the Dixmont hills or Cobscook or Casco Bay. Or more synthetically, in the stars, which after all are the same stars, in virtually the same locations, that all human beings have seen in all times. The white settlers brought Christianity, Platonic philosophy, and science into these woods. This is how the stars came to Maine.

* * *

One winter day I set off through the crusty snow in the Troy woods behind my house with no particular destination. I eventually drifted toward the big round hill south of the house, and when I got my bearing it turned out there were two hills, but that comes later.

At first I legged it down the driveway away from the hill and cut northeasterly into the woods toward a logging road off the main road. I think from time to time loggers drift fecklessly onto the nine acres described in my property deed, but they don't cut enough trees for me to bother the courts about them, and in any case I wasn't thinking about that. It was cold, not more than 10 degrees (Fahrenheit), and the wind was blowing hard through the treetops, swaying the big spruces and firs, and rattling the bare branches of maples and birches. During the summer the loggers had set up a little camp in a clearing, and their rotten-framed couch and the empty cable-spool they used for a table were covered with icy snow. A few fat, limbed maples and oaks were stacked nearby bare of snow, which meant the loggers had been there within the last two weeks. A lot of boot tracks frozen in the snow led away from the clearing but mainly disappeared within a hundred yards or so, as though vanishing miraculously into the lap of the woods.

I turned southeasterly and walked along their road with my hands shoved in my jacket pockets, watching the skidder tracks and small spruces along the sides. There were other tracks. A cat had crossed the trail, and one or several rabbits had hopped across the snow when it was fresh, streaking the surface and poking their little claws down in. One had stopped in the middle of the trail and left ten or twelve tracks in a chaotic pattern, then hopped away. Deer hooves had driven deep into the snow and left hard double-cleft shafts. Raccoon paws. The boot tracks disappeared. Wind swept through the treetops.

Then I noticed some big round tracks which I at first thought were the remains of bootprints, slightly decayed when warm air melted the snow surface. But as I walked on beside them, I saw they were too round for a human foot. I bent down to look at some that were better preserved than others. Bluish snow frozen into big ice-crystals after superficial melting and then more cold.

These tracks definitely had claw marks. I wasn't sure what I was seeing. I was thinking how a boy whose father had given him a rifle and taught him the rudiments of tracking would be able to glance at these tracks and say what animal had made them. An ancient Penobscot, subject to higher laws yet, would know without knowing. I stood up and kept walking and watching.

Then I realized: they were bear tracks. They measured (allowing for the meltdown in the impressions and eyeball estimation) about four and seven inches in length and width, something big. It's unusual to actually see a bear, especially in the more civilized woods of central Maine, and especially when you're not deliberately hunting one. In the northeast most bears are in hibernation by December but sometimes continue feeding through early winter when there's enough food. Most of the previous three months had been unusually warm and probably plentiful, even though this day was cold. In the moments when I understood whose trail I was tracing, my heart sped up and I became more alert to the creaking tree trunks and blue jays squalling intermittently. I was wondering what this bear would look like if I surprised her in the brush by the logging path. Her legs down, back parallel to the ground, gazing directly at me as if I were due north. Hazy, dubious gaze. Maybe she'd run.

This wasn't going to happen, though. The bear walked through here days ago. I followed the tracks to where the road divided in a big open area and cut to the right, southwesterly, through red osier and shriveled burdock into big firs. The bear continued along this trail. Sharp deer hoofs had cleft the snow more recently. By this time the road was making its way up the hill, and I labored in the snow.

Now what I was doing was following this bear, who at that moment was no doubt miles away, probably behind me to the north where towns and houses are scarcer. I was following the tracks more or less southwesterly. I might have hiked out a huge four-sided trapezoid if I'd had days of free time, provisions, and gear, and free of earthly encumbrances could follow the tracks as far as possible. Maybe along the way I could pick up the advice of a guide to show me how to track a bear. I might track down, in my fantasy, a bear, but surely not this particular bear. It wouldn't matter at all in the end because such a journey would be powered by the desire to locate a bear, not the bear. To come face to face with bear.

But although the idea and force of the bear is deep in my psyche the way no other animal is, the experience of a bear in the woods is not. This is my place, but these are not my woods. These are Abenaki woods, no matter what the courts say. I track bear in my imagination only, and if I ever come face to face with a living, growling black bear in the settled boreal woods of central Maine, it will be by sheer chance, or fate. Anyway by this time the bear was light-years away.

What I was up against was the hill. The bear tracks led me where I hadn't known I was going. My boots cracked open the snow crust over and over along the logging road, and my prints angled along beside the bear's. Soon I was walking toward the hilltop rather than toward the animal, and when this crossed my mind the bear tracks turned off the trail and disappeared in the hemlocks. Back in the direction of the house, actually. I bent southerly with the trail thinking I would climb the hill. A big deer had stepped across the clearing and scratched over a raccoon's prints, the cloven hoof was outlined clear and deep in the snow.

The logging road ended—or rather, tailed off in the trees instead of halting in a civilized and definite dead end—near the top of the hill. But as I got there and my bearings cleared a little, I saw that the hilltop I meant to reach was actually a second hill farther on. The one on the logging trail was only a fold in front of the larger hill. So I clambered down the steep bank, grazing fallen firs and holding onto bare poplars and birches, to a low, wet, frozen place. Tangles of snowy roots and fallen trunks stretched a few yards across a tiny bog cut between the hills. Although it was bitterly cold, as we say, the ice looked too clear to be thick. I stepped on the arm and leg shaped roots and trunks.

They reminded me of two powerful images from Maine literature. One is a scene from Kenneth Roberts' novel *Arundel*, where Benedict Arnold's feisty army is diverted by a treacherous authority into the swamps of northwestern Maine on their way to assault Quebec City. Continental soldiers clambered in the cold across a web of roots, their feet broke through from time to time and got wet, and many of them became sick and died in the wilderness, for mile after misguided mile. The woods are close and tunnel-like.

The other image is of the deadfall behind the doctor's house in Stephen King's novel *Pet Sematary*. The deadfall is like a barrier between the everyday world and an ancient Indian burial ground containing extremely evil spirits. The doctor has to work very hard to make his way over the deadfall to bury his corpses in the cemetery, where they return to life unlike anything they were before.

I picked my way from root to root in the cold December air with these weird images in sharp focus. These stories were more real to me than the bear I never saw. The swampy, frozen gully between the two hills had, I saw, a life of its own. It has been there for tens of thousands of years, eons. A hundred and fifty years ago when these hills were logged flat, cows grazed among the roots and rolled their eyes and switched their fat tails. The farmers ambled through occasionally, making laconic Yankee comments to each other or the cattle. Maybe they planted cow corn on the treeless hillside.

It has been theorized that sound waves never vanish, they just eternally thin. Everything that has ever been said is still being said all around us. Marconi, the inventor of the telegraph, thought it would be possible to build a device that detected old, thin sound waves; with work, he could conceivably tune in to ancient conversations—listen to Jesus giving the Sermon on the Mount, for example.

What would Marconi's device hear in the woods behind my house? Cows lowing, bobcats screaming, wind rushing for millennia through treetops like cosmic background radiation. A few English-speaking voices saying things about the ground that other, far more ancient voices would never have thought of, let alone said—Penobscot voices, the Red Paint people's echoes from seven thousand years ago.

The devil's country. That cosmic background noise must have been frightening. The same Indian voices Stephen King has heard late at night—the woods speak what has been put into them, and a prophet can inspire a dead land just by passing through it, while others' efforts are paralyzed by fright, or bewilderment, or alienation. Roberts and King had their tries and invested a certain fleeting European whisper into the woods, the beginnings of what could be history or shared living memory. But their words will sink into the map, like Arnold's soldiers into the swamp. These woods speak, but they're taciturn. Any meaning our words shape in the woods—Roberts', or King's, or mine—are feeble, distant efforts of people whose ancestors are long lost across the ocean, with no locatable authority anywhere in the vicinity. Any meaning we make seats itself in Greek, Latin, Arthurian, or Celtic worlds, and our words are superimposed here, not adding meaning but possibly subtracting it, a de-meaning of the laconic voices underneath the wind. Penobscots like Joe Polis navigated silent.

I paused on the other side of the ice. The large hill was steeper. I pulled my jacket tighter around my waist and scrambled up a bank of pine needles, frozen sand, and loam. This put me out of breath, and eventually I came onto a gentler slope. The treetops peaked a hundred yards or so above me. I knew I would not be able to see anything at the summit—it was not a bald granite postcard mountaintop with an open view of the surrounding countryside, Knox Ridge to the southwest, Newport to the north, but a foothill of Mount Harris in the Dixmont Hills, a few miles east of me. It was a big round hill covered with trees sixty and seventy feet high. At the summit I would still be in the woods.

As I approached the top I saw two little rises more to the west of south. I wanted to stand at the highest point, as if it were possible to view the universe from some ultimate endpoint like Alkaid. I stopped to look. Tracks dotted the snow and the air seemed colder. The easterly rise seemed higher than the other. As I walked toward it, breaking dead pine branches and stepping carefully into a gully filled with icy bramble and spruce, the wind suddenly picked up and began to rush in the trees over my head.

In big wind the trunks of stiff, cold trees groan like ghosts. All around me the trees were hoarse with muffled, hollow, prolonged warping sounds, like boats creaking at anchor in ocean swells. Twigs and branches snapped louder in the gusts. One snap followed another so it was impossible to gauge the directions or even the distances of the sounds. The maples, which are straight and arrowlike when hemmed in by other trees, swayed like skyscrapers.

The hilltop was creased and rutted, irregular, real Maine topography. I sank my boots into brittle tangles, over hole-like gullies, and around the bare poles of young birches and ash. The wind began to roar. I was at the summit where that northern gale rammed headlong, unobstructed across the treetops. Lower down I was protected by the trees and slope, but here it infiltrated my nylon flight jacket. On the hilltop, things were different. The wind was wicked.

It circled in the trees and blew down inside my jacket. An inner chill cut through me, not cold penetrating in, but something inside my bones pressing out. The hole-like shadows under snow-rocks and bent poplars went askew, as if the dark had folded over in those ragged places, sort of the opposite of snow-blindness. I was not yet in a struggle, but I had the distinct, almost hypnotic impression of being warned off. Something lived here, outside history, or fiction, or memory.

The air was very clear, the cold and the bright blue sky in the treetops, but pale because the winter sun is low and throws as much shadow as light. The trees cracked and groaned. This experience already inhabited my psyche, from Knocknarea. The wind was not everything here because trees blocked it, but it spoke instead, the creaking and a sound like a railing squirrel welled up as if from a bone yard. I went to stand on the highest place anyway, or what I thought was the highest place. The wind forced itself across the maple tops and down my neck.

Westerly, at an odd angle across the northern slope of the hill, the light looked muffled. If I was not so set in my ways I would have seen something extraordinary and frightening. But all I saw were trees and snow, tiny gulches and rags of dead or sleeping plant life with narrow conical displacements of light powdering them, all things you could name and describe. Behind them was what an Abenaki would have seen and what I could neither describe nor therefore make out. Secret, bearlike, close, abnormal, and harsh.

My scientific, naturalist bearings planted me firmly on the place I took to be the top tiptop. If I dared to come there in the evening and peer through the treetops, from where I was standing I would be looking directly at Ursa Major tilted handle-down in the sky, rotating around Polaris. It would be, after all, not the woods I was writing about at all because I see nothing in the woods but trees and tracks, snow, lichens, pine needles, occasionally a rabbit, raccoon, or deer, and never a bear. It would be not the woods but the stars I am writing about because they are so immediately remote. At this time of day, the dipper is handle-up but invisible. I know so many random star facts, and have seen them so often in my telescope, and thought of them in such detail in my own imagination, that even though they reside in places completely deserted to the human psyche, they're sharper in my thoughts and philosophies than the brutality on the hill beside my house. Whose woods these are I think I know, and they are not mine.

I stepped quickly across the snow and skidded back down the slope. In a minute or two the roaring subsided to a distant cosmic whisper. I slipped down over the bare bank and came to the frozen gully between the hills. Somehow the ice seemed thicker and safer than it did going up, but I stepped carefully nevertheless. This was the way back from an ancient place which, whatever it was, had not sold off its holdings to the government. I crossed over.

The slope of the smaller hill was steep but bare and there were easier toeholds and handholds on birch saplings. I came to the top of the smaller hill where the logging road vanished and cut back down the hill in the direction of the house, which was nearly due north. If it was treeless and dark, I would see Polaris like a shining pinpoint exactly over the front door.

On the way I kept an eye open for Christmas trees that my little son and I might cut next year. When I came onto the path that he and I often take on our short hikes, I was amazed to see our own footprints traveling first up and then back down on the snow crust. We had been there several weeks ago to cut our little ragged spruce and drag it home to be made fun of by the rest of the family. Our tracks were in the same deteriorated shape as the bear tracks a half-mile to the east. That bear was long gone.

The Blessing of Snow in Winter

Thank goodness it still snows.

In the week or so after New Year in 2015, it looked like we might be in for a repeat of 2012, when what little snow fell early on failed to cover the ground permanently, and by mid-January the woods and fields were as bare as November. Only worse. When the snow disappeared again in December 2014 and the dead woods were exposed, the same haunting inner shadow started to come over me.

In autumn, cornstalks and bare birches look like the fraying edge of summer. The last of that decaying brown lushness Keats called "mellow fruitfulness" and put side by side with gathering swallows, the dreamlike odor of poppies, and the word "maturing." All of which, if you understand fall in Maine, is superluminally parallel to geese in flight, milkweed silk, and the rustle of leaves on the sidewalk. It has a fullness and dignity available only at that certain late moment. Then winter settles in, snow blanks everything, and ice stills the rest.

But when the same cornfields and deciduous woods stay naked into January, the fullness is gone, and the remains, at least to me, are creepy. Without obliterating snow, it's all laid bare in a way that atheists try not to think about. It's not winter. It's the underside of death.

I can hardly stand to look at the trees. In October, late-afternoon sunlight practically sets the maturing vegetation on fire. But in January, the sun sputters to a high point not even clearing the firs and sends shadows the length of evening across the road at one o'clock. By two, lurid orange-purple light is seeping into chokecherry tangles. Shriveled cornstalks gleam dully in low-angling rags of pallid sun. Maple skeletons writhe on the roadside. Leaves muckfrozen to the ground. Apple trees twisting like forsaken spirits up out of the bog, with a few sickly, unfallen fruits dangling off their limbs like ghosts of Christmas ornaments. A birch that year inexplicably broke and crashed across the chain-link fence at the park. For a while winter did not have the common decency to cover it up.

Mangled, leafless, lifeless, snowless January so completely lost to summer that there appears, from here, no exit.

I start thinking: is this what we've had to look forward to all this time? In Stephen King's story *Revival* the main character discovers that, despite his lifelong denials, there is an afterlife after all, but on the other side are, not flights of angels singing folks to their rest, but unspeakable sufferings toward which everything funnels with Bataan-like efficiency and no escape. Thank god I'm still alive for now, he thinks.

Or are the twisted apple trees and livid cornfields what climate-change suicide looks like, funneling us all down its drain with cosmic efficiency?

Whoa, wait a minute, this is really getting carried away by the perverseness of winter scenery that, thankfully, flickered only a moment that year. Relax. Even if a lack of January snow does have something to do with climate change, we're still alive for now, and didn't have to shovel anything.

It's getting easier and easier to be agnostic about the once and future weather. I could be snowbound in a nutshell and count myself the king of backwoods Troy, were it not that I have bad dreams.

Dark and Reckless (Part 2)

Winter is a dark and reckless thing, even when it's half asleep. By which I mean, of course, what happens when there's a lack of snow cover. At my house in Troy, we have mixed feelings about this.

In one way, we feel like we're escaping execution. Every day the driveway is visible is another day to pretend burial by ice crystal is a month or more away. There can be nippy nights, some minus-single-digit mornings. A daylong flurry around New Year might frost the ground but then vanish in most places, returning the brown and gray of extended Thanksgiving-time. By any standard of dread, a welcome delay of burial.

On the other hand, I remember in younger days, the 1980s to be exact, spending blizzardlike amounts of energy waging mental war against global warming. Winter was clearly—to me, at least—warmer, shorter, and less snowy than when I was a kid in the 1960s. We used to jump off the roof of our ranch house in Cape Elizabeth into snowdrifts. By 1983 or so there appeared to be no longer any such possibility in southern Maine. I remember fretfully watching the newspaper's degree-day reports ascending steadily over each previous year's report. (A "degree day" is a figure indicating how much energy on average is needed for heating buildings—less and less year after year.) It seemed like the environmental apocalypse was gathering to rip us limb from limb. Not one to neglect an opportunity for psychic friction, I resisted it with all my might inside the confines of my own skull.

By the 2010s, it seemed pretty clearly not to have been a phantom of paranoia. Extreme heat waves, floods, tornadoes, and winter storms erupted at record disaster levels in 2011. Glaciers are melting. Droughts have whole climatures by the throat. In several years that decade it was mid-January before there was permanent snow cover in Troy. A huge majority of climate scientists—somewhere in the range of 99 percent, or so I have read and do in part believe it—agree the Earth is warming, and the vast majority of those agree that human activity is a contributor if not the outright cause.

Now this is troubling, and surely bodes some strange eruptions to our Earth. But sometimes I'm not as upset about it as I was when I was thirty. Anxious mental hand-wringing is futile. What can I do about the sea of troubles a warming climate will bring?

While I am well aware that the lack of snow is probably yet another portent, on the other hand Januarys have arrived when I had yet to shovel hardly anything white. I get up and down my luge-run gravel driveway, both by foot and auto, with almost no difficulties. And the blue jays, squirrels, and other neighbors appear to thrive even though it's technically the dead of winter. Is it this respect for simple conveniences that makes calamity, after living so long?

Every year this happens, the meteorologists carefully point out that it has actually snowed; the disjointed part is that it has also melted. By a rough calculation based on local average figures, we get around twenty inches of snow in any given January in our wooded dell off Route 9. But as my wife, Bonnie, frequently observes, even when there's no snow, we still have more than everybody else. When Bangor gets six inches, we get ten. So when the rest of the world is mostly brown and gray, a white New Year's coverlet usually still lurks in our woods.

January

Which is sort of comforting, because the brown and gray in the world beyond our driveway troubles my mind's eye. While there's a barren beauty in November's floral skeletons and the green winking out to tan and brown, two months later the twisting crooks and joints of bare birch and maple branches seem cadaverous, or like they're fighting in some claustrophobic panic. Surely many bushes and shrubs that depend on snow insulation burn in the open cold of January.

My mind just cannot let sleeping winter lie. Despite their conveniences, snowless winters are out of joint.

Deep Freeze

In good old-fashioned winter deep freezes, all you really want to do is get home.

But getting there is like crossing interplanetary space, when the cold really takes hold. On a night highway, headlights from other lanes well up in an ominous glow, icy trees in the median glisten in Transylvanian crystal shells. Are any sparks of life even in them still?

When I left Augusta late one New Year's Eve, my dashboard thermometer registered minus 2 degrees (Fahrenheit) and home seemed eons away. In the blackness of highway space it dropped to minus 5 around the Sidney exit.

A few miles on in Waterville it was minus 7. On the back road through Benton it went to minus 10. The farther out you go, the colder it gets, I thought. And thinking this, I found myself on a kind of intermunicipal trajectory whose map, if you were to draw it, would look like the curving lines traced out by spacecraft making their ways among the asteroids and planets. Cassini on its way to Saturn in the early 2000s. Voyager II bending around Jupiter and past Saturn, then to Uranus, Neptune, and the trans-Neptunian space where it's still speeding outward. New Horizons zeroing in on Pluto and passing it cold.

Colder and colder. The average temperature on nearby Mars is about minus 20. Next out by planetary orbit, Jupiter's average atmospheric temperature is about minus 160. At Saturn, it's around minus 220, Uranus around minus 320, and Neptune around minus 340.

None of this seems too tough, really, compared to the wind chill factors between Augusta and Troy that are minus 451 or so. Just kidding. Sort of. The coldest temperature ever recorded on Earth was surprisingly not in Maine, but in Antarctica in July 1983, minus 128.6 F. A satellite measured minus 135.3 on July 31, 2013, also in Antarctica.

Past a certain point, though, it all gets a little squirrelly, in a counterintuitive sort of way. The daytime surface temperature of dwarf planet Pluto is thought to be around minus 380, while its moon Charon's daytime surface temperature may be around minus 370, a bit warmer. So to speak. Neptune's nitrogen-geysering moon Triton, however, which most of the time is quite a bit closer to the sun than Pluto and Charon, has an average temperature of around minus 391—colder than Pluto.

What are the reasons for this warmup over distance, it takes too many words to disentangle here, and no one is absolutely certain of the reasons anyway, at least not in the way everyone is certain it's warmer near the sun because the sun is hot. I mean, something happens out there that is not consistent with the logic of the physical senses. Pluto is usually nine or ten Earth-sun distances farther from the sun than Triton—yet warmer.

But I was going to say, before these facts broke in about the planets (and without even mentioning an ice storm), that when I reached Unity that night the dashboard thermometer said minus 12, colder than Benton, if that's possible. Eastbound on Route 9, that temperature held pretty steady until I actually passed into the intermunicipal space of Troy. At that point the temperature started to rise. Just like at Pluto.

I turned carefully down the driveway and drove through the little tunnel of birches bent from right to left out of the straighter, darker firs. The headlights crazed the enamel of ice that bowed them down in the snowbound bramble. Eight below and glittering in the moonlight. Up the grade to the other end of the driveway and into the slip beside the back door. Seven below.

Over the snow-crust to the steps, with the certain knowledge that inside the house, after this commute measured eventually in astronomical units, there was a spark of warmth. And sure enough, inside, wood and oil and other kinds of burning made a wealth of heat. The house was night-quiet.

Somehow Pluto was warmer than Benton. I wonder what's going on inside it. Its name means the god of the underworld because different kinds of riches (in Greek: πλουτοσ, ploutos) well up from underground. The New Horizons spacecraft, by my calculation, was by then about the same distance from Pluto that Unity is from Troy. It was destined to take some measurements to reveal some of what's in there, like sparks of light in deep space.

The Implicate Order of Winter

In between shoveling and taking increasingly frequent breaks to observe the results, my mind wanders even more now than it used to on otherwise bleak January days. And when this happens, I for some reason get fascinated by the snow.

It's blue.

Now, I noticed this long ago, of course. In between the flakes, so to speak, new snow glows faintly azure. Looking directly at it, you almost can't see it. Like the zodiacal light, it's clearer viewed almost remotely in your peripheral vision.

In the snowbank I was building one morning beside the steps, this blueness unfolded out of the whole pile in the distant glitter of the low, bright January sun.

You might think this skylike radiance results from the same process that produces the blue sky. In the sky, sunlight bounces off molecules and particles in the air and scatters. Since blue light scatters more than red light, the sky looks blue, except directly next to the sun. But in a snowbank, the sunlight is bouncing off the surfaces of snowflake crystals, and so most of the light is scattering more or less evenly and striking your eye as snow white. But ice crystals absorb a certain amount of red light, so there's a bit more blue light angling around, resulting in that azure tinge.

So I was wrong about what process makes snow blue. But then my errant mind, which every January finds the effort to tap into the overwintering processes of consciousness a little more energy-intensive, fixated on what was under the pile.

Every summer a tangle of wild madder grows beside the steps. Its stalks reach up with circlets of eight little leaves out of a sprawling base. Its tiny white flowers spray out like galaxies. Right there in front of me, underneath three feet of sky-blue snow in 4-degree January cold, was the wild madder invisibly waiting for spring.

Wild madder is a bedstraw and an annual, meaning it regenerates from seeds every May. The seed will germinate when three elements—water, oxygen, and heat—stir the tiny plant parts inside. Then they'll unfold into light, which is the fourth element necessary to grow. All seeds, whether they're bedstraws, hemlocks, or grasses, operate the same way.

In one theory of how the physical world works, processes like this are described as being enfolded in the matter. Germination is an order enfolded in the seed. In the right conditions, that order unfolds into a plant. In the theory, which was generated by quantum physicist David Bohm, this process is an example of an "implicate order"; the plant itself, which is a very orderly structure of stalks, leaves, and flowers despite its wild-looking madderness, is an "explicate order."

The natural world as we see it is the visible, explicate order of things. The processes that give rise to the natural world are invisible, implicate orders. Within the seed that will become the wild madder plant is the natural order that is common to all plants. Within the blue light that grazes your eye from inside a new snowbank are the quantum properties of electromagnetic energy—an implicate order. Snow crystals themselves grow by the billions according to an implicate order. When the wild madder unfolds this summer, an implicate order inside will convert light into food.

The implicate order is not simply any process you can't see; it's an order enfolded into what you *can* see. "In the implicate order," Bohm wrote, "the totality of existence is enfolded within each region of space (and time)." The whole universe, in other words, is an unfoldment of implicate order.

Pausing from my snow-shoveling for a few moments in my little region of January cold, I caught a glimpse of blue sky among the ice crystals and saw underneath it to summer, patiently waiting there beside the step to unfold.

Psychology of a Snow Grouch

By the end of January, I'm pretty much sick of shoveling snow.

I know, I know. Snowfalls are still usual in central Maine. But just as one snowflake does not make a storm, one winter does not make a grouch. They accumulate.

When you're a kid, snow has all kinds of good connotations. There's the possibility of no school, of smooth sledding, of vanquishing your little brother with snowballs. Later, snow and ice become a chance to prove your New England hardihood, so you drive your car around without snow tires, help hapless motorists stuck sideways in the road, wear your boots unlaced (or in recent years, Bermuda shorts with no boots at all—I wonder what percentage of those guys survive past March), and hurl shovelfuls of heavy snow like a machine left and right around driveways.

After forty-something adult winters, though—I can't exactly remember back that far—snow ceases to be a lark and starts to look like the subject for a grudge. I get so tired of winter I become a torture myself. How many hours have I spent excavating paths from house to car to Shed? Paths that, for one reason or another, disappear anyway. In recent winters my habit after snowstorms has become to clomp out onto the deck, take up the shovel, and before the first thrust, calculate how soon this might melt. Which it is going to do sometime, so why not wait it out?

The trouble is, it won't disappear anytime soon (see following analysis) and my driveway twists up and down one of those Escher-like multi-incline Maine topographies which—when the temperature rises just enough to create a surface glaze—turns into a five-dimensional luge run. If I leave the snow to melt, we cannot access the cars except by figure skate. And all I have are hockey skates. So I grunt, and shovel. Then fling sand up and down the driveway.

I stop making regular excursions to the bird feeders when the 15-15 guideline kicks in: when the wind chill is less than 15 degrees and/or the accumulated snow depth is greater than 15 inches. This historically takes effect sometime in early January (with unsettling recent exceptions) and remains through March, or lately, April. March is nowhere near as close as it seems, by the way. Complicated thermopsychal calculations developed by me over a long accumulation of frigid January mornings reveal that each 1 unit of time according to a winter calendar is equivalent to 18 units of time in your mental experience of it. So for example, the thermopsychal distance from January 1 to April 1 is 4½ years. In mid-January, spring will be here in approximately four years.*

At one point, one of the 15s allowed me to scrape together enough sense of adventure to tack out to the bird feeders. Boots were sinking in only about six inches, so I did not have to set up a base camp (i.e., shovel snow) in a snowbank to get there. The chickadees took up positions in the sumac branches inches from my head and scolded the hell out of me—*hurry up-up-up-up-up*! In a few minutes came nuthatches, who bully chickadees. Afterward blue jays, who bully everybody.

The word "grouch" was first used by Americans around the year 1900, according to the *Chambers Dictionary of Etymology*. The philologists are not sure how we came by it. It seems

---

* The thermopsychal effect is inverse in summer, when each 18 units of calendric time are equivalent to 1 unit of thermopsychal time. So the thermopsychal distance from June 1 to Sept. 1 is 5.1 days.

to be related to the archaic English word *grutch*, meaning "to complain" and (according to the OED) "grudge" and "to make a jarring or grating sound." Chaucer's Middle English word *grucchen* (which froze onto my imagination when I first encountered it one February about forty-five years ago) means "grumble." This means grouching was an activity before it became a person. My guess is the first grouch lived in New England and had shoveled a lot of freaking snow.

February

Snow

Brute Neighbors in Troy (Part 1)

For a few days early in one February the driveway ice was covered about an inch deep with dry, powdery snow. As I made my lumbering way up through its red and blue sunglints on one of those mornings to retrieve the newspaper, I saw my footprints were tracking neatly along beside all kinds of other feet. One of the cats had padded across to the garage an hour or two before. Farther down under the ancient white pine a rabbit had hopped hind feet by front from woods right to woods left. Sometime before or after that, a mouse or vole, or some such miniature, left its little wispy slips in the dusty snow. Wild turkeys had come through, three-toed. Deer had driven in their cloven hoofs on their regular route. Echoes in the snow.

Before the brook, somebody bigger than a house cat had padded from the east, making a straight line of slightly offset marks for forty or fifty feet, then turned into the woods west. Who was it. I took some photos and shipped them along by invisible electronic channels to some outdoorsmen who know far more about this than I do. A little debate ensued about whether it might have been a coyote, or a red fox, or finally, most likely, a bobcat.

Bobcat!

"It is remarkable," Thoreau says in *Walden*, "how many creatures live wild and free though secret in the woods."

Secret, in there, and dangerous. A few days later there were tracks swollen in the slush that almost certainly belonged to a coyote that had made its way along the side of the road by the mailbox. Meanwhile the neighbors' wildlife camera filmed a blur they thought was probably a fisher. I've followed bear tracks up the hills in back.

Echoes of a whole other world inside the woods. Sometimes you can hear them. Coyotes howling at night, leaving what amount to tracks on the air. The roaring snorts of what must have been a moose just west of the back door. A screech so alien I can't even guess who it was out there in the dark.

On the outskirts of the backyard on the tree line, in the snow-frozen wreck of a February afternoon, we watched three deer browse bare birch and fallen firs. They moved in silence like apparitions seen through the back door glass, darkly. Later I saw their scat and knew they were real.

I wonder if winter is like a long, hunger-ridden dream to unseen coyotes and barely real deer. In the stove-warm living room, you can only imagine.

"A transient acquaintance with any phenomenon is not sufficient to make it completely the subject of your muse," Thoreau wrote more than one hundred sixty Februarys ago. "You must be so conversant with it as to remember it and be reminded of it long afterward, while it lies remotely fair and elysian in the horizon, approachable only by the imagination."

In July we forget how naked the woods are in February, and cold. How we stay inside and read, and listen for echoes.

February 13, 1859: "The old ice is covered with a dry, powdery snow about one inch deep, from which, as I walk toward the sun, this perfectly clear, bright afternoon, at 3.30 o'clock, the colors of the rainbow are reflected from a myriad fine facets. It is as if the dust of diamonds and other precious stones were spread all around. The blue and red predominate."

Sometimes even in this early stage of the sun's return you can hear the gargling, wheezing whir of a chickadee sounding its alarm, the first sense of its presence in a long time. Blue jays squalling too, after extended absence.

There are worlds within worlds, to quote another philosopher of reality. Winter's project is to listen, and track them.

Brute Neighbors in Troy (Part 2)

One summer night there was a bigger commotion than usual near the bird feeder, which hangs in a birch tree among the firs, hemlocks, cedars, and spruces that rise like a wall fifteen feet from the kitchen. I turned on the light and looked out through the full-length glass of the door. Things were churning out there.

I got a flashlight and waved the beam around in the trees. A raccoon had shinnied up the birch and was trying to unscrew the top of the bird feeder with his deft little hands. Another adult was on the ground with a bunch of little ones patrolling around just off the deck. They all looked up.

They had that healthy-looking silvery gray fur, striped tail. Black nose. Dark-framed eyes, shining. The one in the tree stared straight into my eyes. While he was thinking one thing in his little brain, I was trying to divine his thoughts in mine. I wondered if he was wondering: what goes on in there, anyway? I have no idea what raccoons think, of course, but my own theory of mind tells me they do.

They know where comestibles appear, and so their routines brush up against the ones Bonnie and I keep down here in the hollow where we live, mostly like hermits apart from our trudge up the driveway to work. We need to eat, like the raccoon in the bird feeder. And like the skunks who amble around the backyard more or less fearlessly. From time to time they clamber onto the deck and get up on their hind legs to peer in the kitchen door. Does a skunk do a different kind of wondering, behind those opaque black eyes, than a raccoon does?

Through the glass one morning we watched a weasel dart darkly past, maybe thirty feet away in the woods. White and slinky, a black-tipped tail. Dangerous teeth. Deer forage the roughly square-mile territory our nine acres occupy in boreal Waldo County. Their hoof prints mark the snow where they routinely cross the driveway. Sometimes rabbit, turkey, or cat tracks crisscross them and disappear into the woods like stray thoughts. Back up the hill behind the house I've seen bear tracks.

One February night, while I was at work and for all intents and purposes had ceased to exist, Bonnie heard something scratching on the side of the house. At first she swore because she figured it was squirrels again, the little brutes who are a downright nuisance. This was outside the kitchen wall, though, not a place the squirrels usually torment. And it was loud. Really loud. Much bigger than squirrel-loud. It scrabbled across the wall and over the door. She was suddenly a little nervous.

When it sounded like it had cleared the doorway, Bonnie cautiously looked through the glass, but it had climbed too high up to see. I'm guessing it was investigating the Havahart trap I tacked up on the logs and baited with nuts and sunflower seeds to try to catch the squirrels (with no success whatsoever). When she carefully opened the door to look, there, resting on the logs up over the door, was a porcupine, quills and all.

Yikes! Porcupines climb houses? We had seen one recently make its way with purposeful deliberation from the garage—where I've often seen their tranquil Pooh-bear faces and giant brush of spines in my headlights summer nights— across the driveway and under the deck. Later we noted its tracks emerging from under the kitchen side of the deck and trundling into the woods.

It turns out porcupines spend quite a bit of time climbing trees. In summer they eat vegetation such as clover, shrubs, wildflowers, seeds, and in winter when that kind of stuff is scarce they eat the bark of birch, hemlock, and spruce, among others, and according to lore they have a taste for plywood, possibly because of the glue. Where they're abundant they can do considerable damage to commercial timber, though the fact sheets I consulted indicate porcupines don't actually kill trees.

Bonnie had nothing to fear from the quills because she didn't touch them. They don't shoot, as is sometimes believed, but the porcupine fans them out when it feels threatened, and they detach on contact and lock into the predator's flesh. When I was a kid we had a very weak-minded German shepherd who had to be carted off to several overnight stays at the vet after he tried to bite porcupines.

This fellow lowering himself down the door frame had broken his routine to explore the foraging possibilities higher up. Apparently he didn't find anything worth going back for, because we didn't hear him again afterward. (I say him—we have no idea if it was a male or a female.)

The squirrels have repeatedly cleaned out the trap without springing it. The raccoons have emptied the bird feeder. The skunks root around in the compost corral. They're just making their livings. What they're thinking about, I don't know, unless they love the woods for the same reason I do. It's all home to them.

A Spider in Winter

*"Saw today a dark colored spider of the very largest kind on ice—
the Mill pond at E. Woods in Acton."*
—from H.D. Thoreau's journal, December 18, 1855

One late-December afternoon I was looking at the ice-crusted branches of the staghorn sumac in my backyard, when what to my wondering eyes should appear but a spider clambering along a twig in between the frozen clots. It was a longjawed orbweaver, aka family Tetragnathidae with spindly legs and a long body, as spiders go.

What happens to spiders in Maine's harsh winters?

It turns out that the vast majority of spiders—about 85 percent overall—overwinter in soil and leaf litter. The rest tend to die as adults in the fall, depending on each species' life cycle and, to some extent, the climate. Many orbweaving spiders' life cycles involve mating in summer and laying eggs in fall, with the adults dying after one season. Among others, many overwinter in subadult or juvenile stages, and in some cases as adults, a few of which actually reproduce during winter. Overwintering spiders are characterized as winter-active or winter-inactive.

Winter-inactive spiders find shelter in places such as leaf litter or fallen logs, and enter diapause, a sort of hibernation state in which they draw their legs in and go stiff, with their metabolisms greatly reduced, until warmer spring weather reactivates them. Winter-active spiders, on the other hand, continue to move around and feed at temperatures below freezing (32 degrees Fahrenheit). Studies in western Canada found that the most abundant winter-active species were predominantly wolf spiders, followed by sheetweb weavers (primarily in a subfamily of dwarf sheetweavers), and some species of crab spiders. Other spiders known to be winter-active under litter and snow are sac spiders and long-jawed orbweavers like the one I saw in my icy sumac.

Some of the best protection for overwintering spiders is underneath the snow pack, which works as a kind of insulation. About eight inches of snow can maintain a fairly constant temperature of roughly 27 F underneath, even when the air temperature is as low as minus 30. Snow can keep a spider's leaf-litter shelter at around the freezing point, and researchers have found that air temperatures as low as minus 40 hardly affect spiders under snow.

Another way spiders withstand the cold is through a physiological adaptation called supercooling. The mechanism is not well understood, but the hemolymph (or circulating fluid analogous to mammalian blood) in some spiders contains glycerol, which in association with certain proteins acts like an antifreeze. Supercooling among spiders has been observed effective at 18 F and down to 10 F; some species of garden spider (an orbweaver) have resisted freezing to death even when exposed to cold as low as minus 4, and the eggs of some cobweb weavers have remained viable down to minus 40. Spiders in colder climates tend to have lower supercooling points than those in warmer areas; some species of thin-legged wolf spiders in Canada are hardier in deep cold than members of the same species in Great Britain.

It's thought that for some winter-active spiders, starvation could actually be a survival mechanism. As the temperature dips toward the supercooling point, some spiders may cease eating because the hemolymph of some prey may not contain cryoprotectants and thus the meal can freeze in the spider's gut. So it's in the interest of the spider not to ingest freezable food in deep cold.

Some spiders ensure their species' survival by laying eggs in late summer or fall and bundling them into a silk cocoon to overwinter and then hatch in spring. Some spider eggs can remain viable in temperatures as low as minus 11 F. Cross spiders (an orbweaver) mate during late summer, and the female then spins a cocoon and lays her eggs in it. Within a few days of laying the eggs, she dies, and the spiderlings hatch out in the spring. Black-and-yellow garden spiders and barn spiders follow a similar routine, though I watched a mother barn spider stand guard over her egg sac for more than a month before she finally died in early December.

In our part of the world, the fairly common goldenrod crab spiders lay their eggs in late summer. The spiderlings hatch out in the fall, overwinter in the ground, go through a molt in May, and then spend the one full summer of their lives eating bees and producing the next batch of spiderlings.

Some species of wolf spiders mate in late summer, and then the females overwinter in ground shelters or in a building, some remaining winter-active. They spin a cocoon for their eggs in May, and the little spiders hatch in early to midsummer. The spiderlings mature relatively slowly, overwintering as subadults and then reaching adulthood the next summer. Males of these wolf spider species don't seem to live beyond the first summer of their maturity, but females may live several years.

Spiders are so well-adapted in general to the cold that some spiders' life cycles can be longer in colder climates—in some cases extending to up to four years beyond characteristic one- or two-year cycles.

Winter mortality rates are surprisingly low for spiders. Adaptability is the name of their game. And nature's.

Trees in Winter

The older you get, the more the cold takes the starch out of you. Every winter you have to work a little harder to get your mind right to survive. Not only do you have to break out the sweaters and boots, but you have to revise your whole way of thinking about the back steps, the car windshield, and how quickly you can get your bones to cooperate with climbing out of bed.

You even have to adjust your attitude to the scenery. In summer the birch, maple, and ash trees are full-looking and green, the chickadees vanish in the leaves, and you get used to a feeling of abundance. But in winter, the trees look like skeletons, the nuthatches walk around weirdly upside down on the gray trunks, and you're face to face with the fact that the world has given up the ghost. Even sounds die in snow.

In winter, deciduous trees like oak, maple, and beech—the ones whose leaves fall off—have ceased gathering light for food and are as good as dead. Since the leaves are gone, no light is being absorbed by chlorophyll to provide energy for turning carbon dioxide and water into oxygen and carbohydrates. The sap has receded to the roots in the underworld, and the bark, trunk, and branches seem lifeless. They're said to be dormant, which literally means "sleeping" from French, and biologically means having suspended operations.

But even during the deepest sleep in nature, things are happening. Pines, spruces, and hemlocks—the evergreens—still slowly photosynthesize food through their needles. And the deciduous trees look frozen, but aren't. They protect themselves from the cold through a process called "super-cooling." In the fall, as the leaves begin to senesce, or grow old, the sugars the tree produces go unused and instead are stored. When winter descends, the sugars act like antifreeze and can prevent ice crystals from forming in temperatures as low as minus 40 Fahrenheit.

North of us where it gets colder still, an invisible system called "intracellular dehydration" kicks in for some trees. Water freezes, not inside the living cells, but in the spaces between them; in the process, moisture is sucked from the cells, which, being dry, can't freeze, and so the living tissue remains intact even in the dead of winter.

By about mid-March (at least historically), when sunlight increases and the days get warmer, the sap returns from underground. The buds, which are prepared before winter comes, start to generate new leaves for collecting light and producing carbohydrates again.

The amazing thing about this cycle is its inevitability. It never fails. The maple by my driveway seems lifeless from December on, but then in March, or lately, April, its buds redden and it lives again.

There's no way to know how, or if, trees concern themselves with all this, although recent research suggests plants have some kind of sentience—altogether unlike ours, no doubt. Or maybe, not completely unlike. They seem to have a disposition in their bones, as it were, to cooperate with winter. They seem to gain from it. You'd swear that maple loves its appointed times.

Thoreau and the Lichens

Henry David Thoreau, as we all know, spent a lot of time poking around in woods and fields. He spent hours pretty much every day observing nature in its smallest details. Then he'd go home and write up the day's findings in his journals, which amount to thousands of pages and which many scholars say are his masterpiece.

Many of his writings, like *Walden* and a book unpublished in his lifetime titled *Wild Fruits*, a natural history of the plant life around Concord, Massachusetts, are made at least partly from his journals. So reading the journals, even in unedited form (the scholars have been working on them for more than forty years and are still nowhere near finished), is sort of like walking the woods with Thoreau himself. His meticulous observations and ruminations are fascinating.

So I was poking around in these journal-woods during some gray days awhile back when I noticed a strange phrase repeated in a number of places. "It is a lichen day," he writes on February 5, 1853. The same sentence again on January 7, 1855. Weeks later, on January 22 and February 16: "It is a good lichen day." And on December 6, 1859: "It is somewhat of a lichen day." Obviously he's looking at lichens. But what is "a lichen day"?

Lichens aren't really like any other being in the woods. They're a symbiotic combination of two separate organisms, a fungus and an alga—which are as different from each other as both are from plants—growing together as one being. They come in three basic forms: foliose, which includes leaf-like structures with two distinct sides; fruticose, which includes hairlike and mushroomlike structures; and crustose, forming crusts over a surface. They grow on trees, stone, soil, houses, old picnic tables, and other substrate surfaces. They absorb water and nutrients directly from the air and substrate. They go dormant in conditions of insufficient moisture, and turn back on to photosynthesize when conditions are right again. Some individual lichens are estimated to be thousands of years old. They tend to turn gray when dry, and color in moisture. This is the clue.

Thoreau's lichen days are usually damp, often in the middle of a winter thaw. "There is a low mist in the woods," he writes on December 31, 1851. "It is a good day to study lichens." Years later on February 7, 1859: "I see the sulphur lichens on the rails brightening with the moisture. ... A little moisture, a fog, or rain, or melted snow makes [a lichenist's] wilderness to blossom like the rose." A good lichen day is a moist day, when the lichens are actively photosynthesizing and their colors and textures are on full display, edifying the eyes: "A lichenist fats where others starve. His provender never fails."

The spirit is fed by beauty found even in the most unlikely, rudest-seeming form on "the barest rocks." (In fact, in March 1854 he tells us he literally "boiled a handful of rocktripe (*Umbilicaria Muhlen-bergii*) ... for more than an hour. It produced a black pulp, looking somewhat like boiled tea leaves, and was insipid like rice or starch." Like his tries at eating a woodchuck and drinking tea made from arbor vitae, so much for that experiment.)

The scholar Hans Bergmann, who has been exploring references to lichens in Thoreau's journals, noted in a talk awhile back that "Thoreau did not know, as no lichenologist of his time knew, anything about the dual nature of the lichen. ... The lichen is itself a miniature ecosystem."

If he'd lived a few more years to find out the biological truth about lichens, Thoreau would no doubt have found it a profoundly illuminating natural fact. He was among the original naturalists to think of the whole Earth, not just his own woods, as one ecologically integrated process of processes. The whole Earth, he might have said, is a lichen, every part of it growing together as one being.

"The earth I tread on is not a dead inert mass," he goes on to say December 31. "It is a body—has a spirit—is organic—and fluid to the influence of its spirit—and to whatever particle of that spirit that is in me."

Those lichen words lit up my gray February days a hundred and seventy years later. It is a world of words to the end of it, to paraphrase another poet. A whole lichenlike ecology of words.

A Mind of Snow and Ice

In February, it's easy to believe that twenty-five thousand years ago, our part of the world was covered by ice a mile and a half deep. So much ice that the ocean was frozen over offshore about a hundred and eighty miles. So much ice that its weight depressed the earth surface to hundreds of feet beneath where it is today.

In February, I do not find any of this hard to believe. In fact if I had to guess, I'd say that the ice now covering the lower part of my driveway is a remnant of that Laurentide Ice Sheet which never completely melted after things started warming up around fifteen thousand years ago.

The ancient cold cycle starts around the autumn equinox in September, as the Earth's tilt and orbit move us out of the sun's direct rays and arctic air filters south. The cold usually gets its real grip during December, or used to anyway. Floating around in the frigid air are grains of dust and pollen or even bacteria, around which water vapor condenses and freezes into crystals.

How and why water freezes, exactly, is not well understood. Each snow crystal forms in one of seven latticed patterns: plates, stellars, needles, columns, capped columns, spatial dendrites, and irregulars. Apparently, as the International Commission on Snow and Ice concluded in 1951, no two snow crystals are exactly alike. A cubic foot of snow can contain ten million unique flakes. At the time of the glacier there were approximately seven thousand seven hundred twenty-seven million million cubic feet of ice crushing Maine.

And that's not counting the mountains plowed up at the top of my driveway this month.

Thirteen different kinds of ice can exist chemically: Ice I through Ice XIII. But we see only forms of Ice I because earthly temperature and pressure conditions don't allow the other structures to grow naturally. Here in the Great White Northeast we give names to the many manifestations of Ice I that appear here from October, or even in olden times September to … whenever.

Take simple frost, for example. One kind is rime, which forms when water droplets freeze on contact with an object such as a blade of grass. Its cousin, hoarfrost, grows outward directly from vapor to solid.

A thin coating of ice is called verglas; black ice on the road is apparently a kind of verglas. Plate ice forms on still water (e.g., the condition of my lower driveway three to four years per winter month). An ice lens is a buildup of lens-shaped ice in soil; it causes frost heaves. Pingo ice is a huge ice lens underneath a pond, pushing up a hill of ice. Pack ice is long-lasting ice cover over ocean water; an ice floe is a chunk of pack ice moving with a current. An iceberg is a renegade piece of a glacier.

Sometimes, especially in March when winter can't make up its mind whether to further overstay its welcome or depart, you see patches of ground with little frozen points sticking up; this is called pipkrake, or needle ice. Decayed ice with little candlelike protrusions is called candled ice. Fine, small, needle-like shapes and thin, flat, circular plates of ice suspended in water are called frazil.

And then there are the forms of ice we categorize as "snow." There's powder, dusting, and hardpack. There's snow mixed with rain; sleet (defined by the National Weather Service as ice pellets of frozen or mostly frozen or refrozen or partially melted raindrops); freezing rain (falling as liquid, then freezing on impact); and hail (ice falling out of the sky!).

There are kinds of snow you sometimes see but don't name, probably because they're not dangerous. Snow that's partly melted and refrozen and acts like ball-bearings is called corn snow. Little balls of snow about the size of BBs that you sometimes see bouncing around the ground are called graupel. Graupel is different from snow pellets, which are smaller, but larger than snow grains.

Except in the delusive psychic state that, every winter, seems to glacialize all of time into one eternal span euphemistically called "February," we do not have to worry about firn, which is glacial snow that survives at least one season (though in my driveway it seems to be having a shot at it every April) or neve, enormously compacted ice underneath the glacier. In Maine we also do not have permafrost, which is soil that stays frozen for more than two years in a row.

And these are just manifestations of Ice I. What state of mind would we be in if we had to grapple with the graupel of alien ice forms from Ice II, Ice III, or Ice IX?

According to *The Changing Nature of the Maine Woods*, a fascinating book by University of Maine at Farmington professor Andrew Barton, over the past one hundred thousand years it has been deep cold here for tens of thousands of years longer than it's been warm. This is not hard to believe. It sounds like the macrotime version of one Maine seasonal round: cold, colder, cold, and the Fourth of July. And July itself, as far as I can tell, is just a legend at this point.

An Inconvenient Winter

When I got up on a Monday morning late in February of 2015, I was bone-wearied to note that more snow had fallen on already a lot of snow. Two days later on Wednesday, four or five more inches had re-occupied the cleared part of the deck.

It was déja vu all over again. On Monday I brushed the new inch or two off the car. Then I was flinging it with my Big Shovel out of the snow tunnel that leads from the cars (somewhere under there, gravel lies) onto the plow pile higher than the deck.

I cleared this same tunnel again Wednesday. On the Sunday I had also moved a six or eight inch ridge of chunks and slabs that was plowed up in front of the cars for the whateverth time in the past about five weeks. Bonnie and I again knocked down the snow-volcano mounds surrounding the top of the driveway to clear the view of calamitous Route 9. Back-wrenching work I also did on the Saturday. Re-dug the trench to the basement door. Re-dug a trench halfway along the back of the house so the next oil delivery will take place. Spent most of the Friday afternoon on the garage roof. At least a ton of snow went over the edge.

Shovel, shovel, shovel. If this much snow (I start thinking while I'm shoveling) with food and warmth behind the next wall feels like this, what could the Donner Party possibly have felt like?

A NOAA snowfall map shows that the middle swath of Waldo County, where Troy is located, had forty to fifty inches of snow on the ground that Tuesday between Monday's dusting and Wednesday's four or five. The eastern and western edges of the county were in the thirty-five to forty inch range. The forty to fifty range extended into the middle of Penobscot County; thirty-five to forty extended west toward Waterville and east into Hancock County. The lower half of Washington County was in the forty to fifty-plus range.

So on Monday, this one or two inches seemed simultaneously negligible and infuriating, depending on your frame of reference. Everybody from here to Wherever, Mass., Vt., N.Y.—name your favorite snowbound state—knew, that February, what I'm talking about.

Cold, too. An Accuweather chart of high and low temperatures in Troy for February 2015 indicated the high made it into the 20s (Fahrenheit) only seven times from February 1 to 22, and ten days reached 16 or less. The historical average highs for February days are right around 30 for most of the month.

But despite the relentless chill in the morning kitchen, the cold was my ally. It kept the snow from turning into cement. Meaning that light and fluffy is fairly easy to knock off cars, skim from the deck, and push off the garage roof.

Not convenient, mind you. My back hurt all day Sunday and took on a sort of aggrieved disposition again Monday. But if it wasn't for an average daily temperature of something like (just eyeballing the chart, here) 8 degrees, heavy sticky snow could well have extended my discomforts into a range of medical requirements outstripping ibuprofen, ice packs, and adolescent griping at the chiropractor's office.

And really, it's been colder than this. Our outside thermometers never registered 20 below zero that year, which they have in plenty of winters past.

An op-ed piece in *The New York Times* in mid-February, headlined "Boston's Winter from Hell," pointed out startling imagery from a city socked with its record eight feet of snow, some of it beautiful, some of it ridiculous. "But for those of us living here," the writer said, "it's not a pretty picture."

No doubt. But then he said something that, from my vantage point, sounded sort of peculiar:

"We [in Boston] are being devastated by a slow-motion natural disaster of historic proportions. The disaster is eerily quiet. There are no floating bodies or vistas of destroyed homes. But there's no denying that this is a catastrophe." He noted the "crushed roofs, burst gutters, destroyed roads and sidewalks, closed schools and businesses, shut down highways, crippled public transit." The unmoved snow made driving complicated and dangerous. Some people couldn't get to work and were losing wages, and some businesses were likewise losing money.

Boston was a mess, there's no doubt about it. Eager to find out about the catastrophic part of all this, I read to the end.

But somehow, the only things that seemed to rise above the range of extremely inconvenient were the collapsed roofs, the lost wages, and possibly some businesses' lost revenue. It did not seem implausible that people might actually lose their homes due to lost pay. But did that actually happen? For the time being, at least—which turned out to be the rest of the winter—that worst-case tribulation was hypothetical.

If lost income and inconvenient driving constitute catastrophe, then how do we characterize, say, the January 2010 earthquake in Haiti or the December 2004 tsunami in Indonesia, in which hundreds of thousands of people were killed and most of the rest had, not trouble getting to work, but their homes destroyed and families wiped out?

I sympathized with the Bostonians that winter, since my own driveway turned into a crater of eight-foot snowbanks and on a couple of days the plow guy couldn't come until morning and we could not get out. Overall, I was glad we weren't in Boston. But catastrophic? I really don't think anybody was joking around in a yeti suit, as we saw in video clips out of Boston, in Port-au-Prince after the earthquake.

If a shit-ton of snow one year seemed like a "catastrophe" to the writer of that op-ed and all the Bostonians and New Yorkers who agreed with it, I wonder how they'll cope with the weather the changing climate has slated for the world in the next few decades, and for their grandchildren and great-grandchildren beyond.

Although it seemed unbelievable at the time, that snow — that unusually huge lot of it—disappeared, as always, into summer. The rising seas, temperatures, droughts, Category 5 hurricanes, and wildfires to come will not.

Winter in New England is, for now, a thing to be cherished, not whined about.

Snowpocalypse

After the first snowpocalypse in the winter of 2015, we were still alive, contrary to TV predictions.

So as we've done for lifetime after lifetime of snowstorms, we got on our boots and gloves, Bonnie somewhere found my old red and black wool hunting jacket which has shrunk to fit her, took a few deep breaths, grumbled, and stepped out onto the deck.

My ruler disappeared into maybe nineteen inches of pure white snow. The two non-garage cars were disguised as one large snow drift. I started throwing stuff off the deck to create a walk-through tunnel. Bonnie brought out a broom, forced her way to the cars, and started knocking snow onto the trail I was trying to blaze toward the spacious Western Lands of the driveway that the plow guy, Jason Reliability Hunt, had already opened during the night.

Good news: this snow was very light. Up here in Hyperborea we call this kind of snow "light and fluffy." A rural legend holds that the Inuit people—who are much more hyperly boreal than us and keep their frozen hydrates around longer—have twenty-three, forty-two, fifty, one hundred words for snow, while we have just a handful.

But according to linguists, English has at least as many names for snow as the Inuit languages do. Inuit is agglutinative—in other words, endings are affixed to root words, enabling the creation of great, long words which in English would be entire sentences. But in fact, the Inuit refer to snow, ice, slush, and so on using roughly the same number of roots we do in English.

To me the word "snow" already does excellent work. It's old. It incarnated in its present form around 1200 A.D., in early Middle English, a descendant of the Old English word *snaw*, which was cognate with other Germanic words such as Old Frisian *sne*—all referring to precipitation frozen in crystals.

Old Icelandic has *snjor*, Old Irish *snechtae*, Latin *nix* (compare French *neige*), and similarly all the way back through Old Slavic, Old Prussian, and Avestan (an ancient Iranian language) words. In the beginning was the word **sneighw*, as philologists reckon the original Indo-European language had it. Most languages spoken in Europe and South Asia grew from Indo-European, which was in use around seven thousand years ago.

So much snow.

After moving mountains by the garage, we crunched down the driveway to the narrow causeway section over the tree-canopied brook. At that point the grade gets much steeper and a hundred feet farther up disgorges the driveway onto life-and-death Route 9.

When the top is slippery, steering onto the road without getting annihilated by a pickup truck going three-quarters the speed of light is tricky. A few times in the past twenty-five years, we've made it most of the way up the slippery slope only to spot a vehicle missiling down the hill from the direction of Troy Union Church and had to stop, only to start sliding back down.

So our mission, whether we choose to accept it or not, is to scrape snow and ice on the launch pavement that might impede the progress of the front wheels.

The preparatory grouching on this mission anticipates how high the snowplows have banked up the mouth of the driveway since Jason cleared it. After the previous day's breaking of the snowfall seventh seal, we saw as we neared the summit that the bank of gray, replowed snow was little more than a flat rise. Our hearts leaped up. We scraped what we could down to the gravel. Some of it was already packing itself—by a process known technically as Driveway Glaciation Magic—into a plate that was no longer loose snow but not yet ice. We chipped some of it out.

Then we heard a rumbling sound. Down the hill barreled a juggernaut, with huge steel wings flinging leftover snow onto snowbanks.

Our hearts sank. All our good work would now have to be repeated.

"Crap," Bonnie said.

"Instant karma's gonna get you," I said, knowing that in her case this never happens.

We stood back. The plow zigged gently around our neighbor's mailbox, then zagged back, headed our way. Angels and ministers of grace defend us! Soon we would be in full-scale excavation samsara.

The driver bent out around our mailbox, then slowed and lifted the great plow wing as he glided past. When he cleared the driveway mouth, the wing dropped back with a clank and continued on its harrowing way.

Hardly believing our good fortune, we waved and thumbs-upped him.

"That was a miracle," I said. "In all these years of undigging my own grave, no snowplow has ever shown any mercy whatsoever."

So with new energy we started knocking down the cone-like mounds that grow like volcanoes on either side of the driveway and obstruct our view of light-speeding vehicles.

After about five minutes, we simultaneously stopped and listened, like two seagulls staring into the wind.

Unbelievably, more rumbling was swelling from the direction of the church. Then, lumbering like a monstrous avenging fury came a second plow truck.

"No way," I said. "There's no possible way we're going to be spared twice in the same lifetime."

We stood back and waited. Snow flew up from the pavement and the wing was pushing the whole snowbank back four more feet. The truck veered out to avoid the mailboxes and then, at the last second, the driver lifted the plow, slowed down, and stopped right in front of us. The window rolled down.

"How you guys doin'?" he shouted from on high.

"Fantastic, now!" Bonnie said. "Thanks!"

He laughed and waved, gunned the engine, and much to our perplexity, started backing up. Then he dropped the plow and blasted the mound on the right into snowballs. He then angled back onto the road, slammed the big wing back down and resumed pushing the rest of the roadside into the woods.

"This is unbelievable," I said. "This isn't just the day after. This is the actual Rapture. We were saved."

We chipped some more holes in the pre-glaciation, grabbed two newspapers in their plastic bags out of the box, and dragged our shovels back down the hill, across the brook of forgetfulness and on to the ancient snow gates flanking the deck.

"I'm never going to speak a bad word about the plows again," Bonnie said. "From now on, only good words."

"Amen," I said.

Mid-February, the Freezing Point, and Beyond

There comes a point in February when it feels like it's been cold for decades. There has been no human activity in my Shed, which has no source of heat, for at least a century. There have never not been snowbanks closing in the driveway. Winter is lost in time.

Usually the loss is foreshadowed in January. In January of 2009, for example, three of our four cars succumbed, frozen solid for a week. (Don't ask how three people ended up with four cars, this is Maine.) It was the third-coldest January on record in Caribou. The bleak scientific prose of the National Weather Service told the whole story: "The average temperature was 2.5 F which was 7 degrees below normal. ... The temperature did not rise above 30 F during the entire month. The lowest January temperature ever recorded at Caribou occurred on the 16th when the temperature fell to –37 F. This was the second coldest temperature on record in Caribou. On the same day, a –50 F reading was measured at the remote Big Black River observation site." This minus 50 is, as of 2020, officially still the coldest temperature ever recorded in Maine. The previous coldest was minus 48 in Van Buren on January 19, 1925. (The day my mother was born in Portland. Not sure what to make of this.)

Six days in January 2009 had record low temperatures for their dates. In Bangor the temperature went subzero eighteen days. Three of those days were record lows. Minus 25 on January 16, minus 19 on January 25, and minus 26 on January 26.

I don't even want to tell you what we were seeing on our kitchen window thermometers in Troy. It's like a dream where you're being chased but your legs won't lift. The cold is inescapable and it slows everything down, including your knee, elbow, and neck joints, and the synapses in your brain. And the older you get, the slower is the slow. There are winter mornings when I do not want to get out of bed and cannot decide if it's because I can't.

Temperature is a measure of molecular activity. Absolute zero is the total absence of heat energy, the temperature at which all molecular motion stops. This is represented by zero degrees on the Kelvin scale. On the Celsius scale it is minus 273. Fahrenheit, minus 459. Cold nearly that deep in January is bearable, somehow. But by February winter has almost obliterated all recollection of "July," some mythic month invented by first-grade teachers to roughly approximate a Santa Claus of weather.

I wonder if cold gets any enhancement for duration, like it does for wind. A minus 18 degrees morning must have a time-chill factor because I remember one when, months into an old-time relentlessly cold winter, with the engine of my station wagon grinding and gasping, the air felt roughly equivalent to minus 458. The car could not move. I could not move. The trees could not move. Nothing was moving. In my yard, all molecular motion had stopped.

I've heard that after minus 50, spit crackles and freezes solid before it hits the ground. The coldest temperature ever recorded in the Northern Hemisphere, according to the World Meteorological Organization, was minus 93.3 F at an automatic weather station smack dab in the middle of Greenland, December 22, 1991. The lowest temperature ever recorded on Earth was minus 135 F in Antarctica in July 2013.

You will note that in Maine, it was summer when that happened.

Somehow it comes as no shock. As far as I can tell, by February winter has taken over the fabric of space-time, and soon the time-chill factor will stop all the cosmic wheels. Winter seems to have become a place from whose bourn no dreamer returns. It puzzles the will. At length comes March, or so I learned when young.

Sometimes a Glacial Notion

Every winter, a sheet of ice covers the bottom of our drive-way. By "driveway" I mean the gravel roadway roughly ten feet wide that bends for one-tenth of a mile down from the road then up to our backyard and may be suitable for use in Olympic luge events. The spot where it crosses the brook is similar to a gravel railroad causeway and too narrow and dangerous for sledding. We drive cars there, though. You have to watch your speed.

By "bottom," I mean a flat, shaded expanse just the house side of the causeway where water comes to rest and freezes solid. This ice at the bottom is believed to be a remnant of the Laurentide Ice Sheet, a glacier that covered Maine, Québec, New York, and who knows where else about twenty-five thousand years ago. To find out if it really is a glacier, a certi-fiable backyard naturalist (me) collects and analyzes data per-taining to the formation, longevity, and potential futures of glaciation, particularly in Troy, Maine. This research seems to resume around February every year.

Does the Troy Driveway Ice Sheet meet the definition of a glacier? Apparently. A glacier, according to the Extreme Ice Survey, is a mass of ice that originates on land. It is usually larger than one hundred square feet. It forms from snow fall-ing in layers; the upper layers of snow compress the lower layers into ice.

These are exactly the conditions at the bottom of the driveway. In November or December, snow falls. If it's suffi-ciently cold, then that snow, under the influence of the so-called wheel-rut factor (which speeds up compression of the snow and causes surface crystals to melt and glaze), soon turns into ice. When it snows again, that second layer covers and presumably compresses the first layer; the third covers the second, and so on. Over the ten thousand to twenty thousand years that elapse between December and March, an ice sheet two to five inches thick forms on the bottom of the driveway. Sometimes a notion crosses my mind that it's suitable for skating. But that's off the track.

EIS also says a glacier is a "year-round mass of ice." Now, by February, it is never remembered for sure whether the ice on the driveway ever wasn't there. It does not seem like it. It seems like it has been there since the beginning of time. Lacking evidence of any such season as "summer" other than in remote legend, it is assumed that winter here is for all intents and purposes perpetual. If so, then the Troy Driveway Ice Sheet is a year-round mass of ice.

There are two kinds of glacier: Alpine, or mountain glaciers, such as those in Alaska, the Alps, and Mount Kilimanjaro, and continental ice sheets, such as those in Antarctica and Greenland. The Troy Driveway Ice Sheet appears to be a mountain glacier because it is somewhat smaller than a polar ice cap and because it appears to flow down the hill and carve valleys in the driveway that appear first in March, proceed well into March proper, continue in March, and go on into that period of March, if not April, when the mud-freeze-mud-freeze cycle, on top of ten thousand years of winter, makes you think you can't really take it anymore.

Good news, though. The Troy Driveway Ice Sheet may disappear. Mountain glaciers worldwide have shrunk markedly in the last forty years, and climatologists predict that at the current rate, they will all be gone within a hundred years. Calving—the breaking off of big chunks of ice at glacier outflow areas—has speeded up. Slabs of ice much larger than any previously known have broken off in recent years. One twice the size of Manhattan fell off Greenland's Petermann Glacier in July 2012.

The melting responsible for disappearing ice is caused by climate change, aka global warming, which is actually happening regardless of how cold it was at your house last night. In the last fifteen years or so, temperatures have risen about 5 degrees Fahrenheit on the Greenland ice sheet. In a heat wave during July 2019, it lost about one hundred ninety-seven billion tons of ice to melt, enough to raise global sea level by about two one-hundredths of an inch. On one day, July 31, 2019, ten billion tons of ice went into the ocean as surface melt. On August 1, 12.5 billion tons of ice were lost.

The warming might be part of a long-term climate cycle, but it is certainly influenced by a steep rise in carbon dioxide levels unknown in the last half-million years. Corresponding to the steep rise is the fact that at the same time, humans have been pouring carbon dioxide and other greenhouse gases into the atmosphere at high rates. It is practically certain the three kinds of rise are linked.

This might all be at worst a shrug—so what?—or at best a chance to make a pot of money, which is the way some optimistic, avaristic politicians see it. The trouble is, there's trouble ahead and trouble behind. If the glaciers keep melting, within a hundred years sea level will rise by about three to four feet. Worldwide, about one hundred million people live within three feet of sea level. One hundred years is headed like a freight train straight at you, me, and everybody else whether we turn two good eyes toward it or not. If in some future future all the glaciers melt, then sea level will rise two hundred feet. Meanwhile, mountain glaciers are melting now. A billion people in Asia alone depend on Himalayan glaciers for water. When their drinking water runs dry and their homes are under the ocean, what will happen?

Take my advice, you're better off not thinking about it. Here in the woods of Troy, the key environmental question is not how billions of people in Europe and Asia will get drinking water, but if or when the ice sheet is going to melt. Those are just two completely different things. That and driving the car over the brook, and along the ice. When you come around the bend, you know it's the end.

Sun Signs

In February, every year, something wonderful happens. Around 5:15 or so one afternoon in mid-month, you suddenly realize: wait a minute, it's not dark. You haven't turned your car headlights on. You can see chickadees from the kitchen window at suppertime. You didn't have to turn on a light this morning.

We all carry around a sort of transcendentalist's faith in the reality of summer, so we know there's more to the world than flickering afternoon shadow. But what's startling is how the sun takes you by surprise. Daylight in late afternoon, a miracle of nature. And of course, careful cosmic arrangements are made for this awakening to occur.

The seasons arise in the first place because the Earth, as it orbits, is tilted on its axis with respect to the sun. That tilt always (for present purposes, anyway) points in the same direction, and so as the Earth circles, the North and South poles pass slowly in and out of the sun's rays. In other words, from around March to around September, the North Pole is tilted toward the sun. The effect we see is the sun higher in the sky, giving us more daylight. Between September and March, the North Pole is tilted away from the sun, and the effect we see is the sun lower in the sky, giving us less daylight, and winter.

An interesting facet of this motion is that the sun does not get higher (from December to June) and lower (June to December) at an even rate. At the two solstices it seems to stall and not get much higher or lower, and at other times it kicks up and gets higher or lower faster day over day than at other times. The best example occurs in the dead of winter. At the winter solstice (around December 21), the North Pole is facing away from the sun, and so for us the sun is at its lowest point in the midday sky. It barely clears the trees here in the Troy backyard. After that, the Northern Hemisphere starts edging back into the sun's rays. The sun starts getting a little higher in the sky each day.

Around January 1, the sun is gaining less than half a degree of altitude in the sky every five days. This is part of our problem in winter—months into the cold and snow, the sun doesn't seem to be getting any brighter. This shadowy winterworld creeps on into January. Around the beginning of February sunset is around forty minutes later than it was a month earlier: the tilt of the Earth is taking us into night's shadows at a pretty steady six or seven minutes later per five days.

But, if you watch the sun around noontime every day, you might notice in early February that it's reaching a higher high point at a faster rate than it was in early January—a whole degree more gained per five days. In mid-February this heightening height kicks into a sort of mad cosmic gear. By the third week in February, the sun is climbing more than 1½ degrees higher in the sky every five days. More than triple the rate in early January. It's around this time that the revelation of sunlight strikes.

The main factor in this quickening gain in altitude, along with the tilt, is that the Earth's orbit around the sun is an ellipse, or slightly flattened circle. So the Earth is closer to the sun at some points and farther away at others. A couple of weeks after the winter solstice, the Earth reaches its closest approach to the sun, and is moving faster than when it's farther away. The Earth's speed, its elliptical path, and the angle of the tilt combine to change the sun's angle in the sky more quickly from day to day. The effect we see is the sun climbing faster and getting brighter. The effect we feel is that liberation from the cave seems imminent.

But as everyone in Maine well knows, we still have March to deal with, and lately, April. The nights stay angry-dog cold. Beshackling blizzards are still possible. But the sun's climb kicks up 2 full degrees of altitude per five days through the spring equinox (around March 21), and sunlight floods the days. In April, as the Earth slows, the frantic climb tapers off. By sometime in May, summer has won. After that it's all June and July.

Beyond the cave is sunlight. It is not a myth.

Winter Driving

"Reality is that which, when you stop believing in it, doesn't go away."
  —Philip K. Dick, "How to Build a Universe That Doesn't Fall Apart Two Days Later"

Long winters ago when my son Jack was still a boy, we were driving along Route 9 to Unity a few days after a snowstorm. The road was bare, mostly, but sub-20-degree cold had been on for what seemed like weeks, and so broken ridges and mini-mesas of ice were cemented on the road like huge barnacles.

You know what that's like. It's not snowing, so as long as you keep your wheels more or less in the tracks where the pavement appears to be bare, you think you can sail along like it's endless summer.

Still, I kept it under fifty for safety's sake because, after all, my seven-year-old son was riding happily beside me.

Funny story, it turned out.

In front of Espositos' pottery shop we were suddenly going sideways. How this could have happened, I didn't know, and how it turned out the way it did, I understand even less. Because coming directly at my driver's side window were a car and a pickup truck. Or rather, my driver's side window was sliding directly at them.

I'm pretty sure my hands and feet switched instantly to skid mode, which after nearly thirty years of driving on two continents I had of course experienced and, as I believed, had the knack of. But I don't actually remember it. The next thing I knew, we were jolting backward into the snowbank in front of the pottery shop.

I looked over at Jack. "Are you OK?" I said. He said yes. His eyes were big, but it was less because he was scared and more because he wanted to scramble out and see what non-Euclidean angles the outside of the car had ended up in. He couldn't open his door because the passenger side was wedged into the snowbank. So he crawled out the driver side after me.

By then the oncoming car and pickup had stopped to see if we survived. Faces were peering at us with real concern, because that's how most people get in dangerous circumstances. Not all, but I'm coming to that. The guy from the car took my arm to help me out and asked if we were OK.

"Yes," I said, "but how?" He shook his head in equal disbelief. How our car could have been sliding directly at his grill in the left lane but then somehow change direction, re-cross the road to the right lane, and plow backward into the other snowbank unscathed, neither of us could remember or even visualize. Twenty years later, I still can't. I don't know how the physics of why we all didn't collide and end up mangled or dead that afternoon worked. It makes you give serious thought to the concept of "angels."

What holds vehicle tires to the surface of the road is friction. It turns out the scientists who study tires on icy pavement refer to two kinds: wet friction and dry friction. Ice normally has a thin layer of liquid water on its surface, which seems counterintuitive, but the liquid is the reason the ice is slippery. Ice in a temperature right around freezing (32 degrees Fahrenheit) and calm air has a relatively thick layer of liquid water (wet) and is very slippery. When the temperature drops below about 25 and down to single digits, that layer of liquid gets thinner, and the ice is less slippery. Less slippery still, if wind is scouring it.

The friction needed to brake the car has three components: the ice and ambient conditions; the tread and composition of the tires; and the vehicle itself. The little black Mazda that long-ago day was only a few years old, and the all-weather tires were sound. But the surface texture of the ice was no doubt smooth after the previous cold week or more of vehicles grinding down the ridges. The temperature that day had come up toward freezing. So we were driving on patches of ice with a pretty low friction coefficient. And it's not unreasonable to think black ice had invisibly formed on what looked like bare road. In front of the pottery shop, the friction let go.

I was thinking about this one weekend twenty or so years later while I was driving along Route 3 in China just before a blizzard. It wasn't actually snowing yet, so there was an illusion of freedom to sail. But in the roadway were ridges of ice, just like in long-ago Unity, and stretches of slightly thawed slush from snow that had fallen earlier. Wet ice. Slippery. I kept my speed around forty-five miles per hour, which is the statistical cusp between crashes and crashes with serious injury or fatality.

Up behind me came a van. Not the most agile of vehicles, especially on ice. Undeterred by, or oblivious to conditions, the driver tailgated me for several miles. On one of the "slower traffic keep right" hills, the right lane was an undisturbed layer of half-frozen slush that undoubtedly concealed shelves of ice with developing layers of liquid water. So I kept to the tracks in the left lane where everybody before me had gone too.

The van driver gunned up the right lane through the slush, got enough momentum to pass me before the right lane ended at the top of the hill, and shoved his arm out the window with his middle finger up as he cut in front of me and took off going at least sixty-five. It seemed like dumb luck when in the next ten minutes I did not come upon a crash.

Of course, I have no idea what that van driver did or didn't understand about icy roads. Don't know whether he was a daredevil risking everybody else's life and well-being just for the hell of shoving recklessness in everybody else's face, or just an ignorant fool with no functioning relationship to the real world of ice, or both. The thing is, those were the only two possibilities I could think of, and I was on the same road with him.

That guy probably votes, I thought. I spent the rest of my drive to Augusta trying to tame a low-grade panic induced by the sudden understanding of how icy the roads had actually gotten by that point.

I don't know what I think about angels, to tell you the truth.

March

Dog of Winter

It Is Still Very Cold

First of all, I am not going to talk about March because it evidently is frequently, of late, getting postponed until April. In the winter of 2014, as late as March 24 and 25 it was 3 degrees outside my kitchen window around 8 a.m., not to mention March 1 and 2 and the bone-chilling nights in between. OK?

And since my mission here is to introduce some ambivalence into the feeling that winter has the capacity to make oblivion seem like a plausible lifestyle alternative, it will just be better not to talk about the first three weeks of March 2014.

Our Australian newsroom colleague said that before he arrived here in December, he heard Mainers are obsessed by the weather, and subsequently discovered it was true. He had not yet seen it, but we also have an irrational love of August, September, and October, when the most gorgeous conditions of air, sun, plant life, and starlight anywhere on Earth overpower all beauty-detecting faculties, and in that stretch we utterly forget about January, February, and March which even in unremarkable years can kill you while you're sleeping.

Winter is a battle against natural treachery. Simply driving the car takes a lion's nerve: on any sunny afternoon black ice can send you suddenly to your account. The walk to the garage is an icy ambush. Cracks in doorframes and window sills are natural gates and alleys for cold air. In the last stages, every precipitation freezes to a vile and loathsome crust at night and melts to curds and creases of muck by day.

That year the cold felt bitterer than usual. Some of us, sensing doomsday near, felt sick at heart from the biting outset in mid-December 2013. Sententious, ineffectual soliloquies warning Mother Nature to lay off were being muttered in wind-bitten parking lots by mid-January. Cabin fever madness hit sooner than usual. It felt like the coldest winter in modern times.

Numbers gleaned from conventional sources corroborated the feeling that it was colder than usual. However, its severity may have been an illusion. I have a backyard theory about why it seemed worse than it really was. Bear with me, there's a method in it.

Facts and figures. The National Weather Service reported winter temperatures in Maine were about 2 degrees (Fahrenheit) below normal for 2014. It was "significantly below average" in December 2013, a little above normal for January, and below average in February.

What does this mean, specifically. In Troy, the average high temperature in January over the previous thirty years was 18 degrees, and the average low about 10. In February, the high is typically 17, low about 7. Now I could not find calculated averages for Troy for 2014, but eyeballing internet graphs and charts, it looks like our temperatures were indeed lurking just beneath those average numbers.

In Portland, the overall average high temperature in January is 34 and in February 36; the average low for both months is 15. A chart for Portland showed the average high in January 2014 was 30 and the average low 13; for February, the average high was 31 and average low 13. So a couple of degrees colder, as the NWS said. In Bangor, January averages 27 and 7, and February, 31 and 10; if I understand the NWS report correctly, Bangor's December 2013 through February 2014 average high was 28 and the average low was 9.

North of us, where it is always colder than we like to think about, Québec City's average January high is 21, and average low is 3; in February it's 23 and 3. Climat–Québec says the overall mean temperature up there was near normal in January 2014, with some higher and lower discrepancies in different parts of the province, and about 2 degrees colder than normal in February, like here.

So while we were fretting in Troy about low temperatures that probably averaged around 8 and 5 degrees, Québec was churning along under lows averaging about 3 and 1.

Meanwhile, up in Labrador City, which is about 9 degrees of latitude north of Troy, the normal high on any given January day is 4, and the low on any given night is minus 17. In February it's not a lot different, 7 and minus 17. An eyeball look at some of their daily numbers for 2014 suggested they had a winter of more or less normally blood-crystallizing temperatures.

So really, it was not as bad as we thought. Now here comes the crackpot theory. It seemed outrageously cold that winter, not because the temperature was hugely lower on average—it obviously wasn't—but because the temperature was low more relentlessly than usual. We could not stop thinking about it from mid-December on. Morning after morning, week after week the thermometer on my deck read single digits or lower, except for a few scattered days in January and February. By mid-February it stopped making up its mind at all and flatlined, coming to its senses only when sunshine shot over the roof and heated the outside component.

Relentless, will-withering cold. There's nothing you can do about its slings and arrows. You just have to get your mind in a nobler place, and bundle up your bodkin.

Sometime, the last shreds and patches of our imprisoning seven-foot snowbanks will disappear. And the crazy obsession with winter transforms, as it always does, into the madness of summer beauty. Until then, the rest, from me, is silence.

A Winter from Which I Am Trying to Awake

Don't get me wrong—I'm not sorry our driveway was not enclosed by eight-foot snowbanks this particular winter (of 2016). The frequent presence of an ice-free walking route from the deck to the garage mitigated the claustrophobia—an infirmity of the brain associated with cabin fever—that perennially haunts January and February afternoons. Was it my imagination, or did the sun not dip as low in the midday sky as usual?

Well, that had to be my imagination. But the effects felt here of the giant El Nino warming in the Pacific Ocean were not imaginary, and I have to tell you, much as I was glad of the general absence of snow, I have found it kind of unsettling and dreamlike at the same time. Kind of creepy, actually.

There have been February days in the past when the temperature reached 50 degrees, of course. But this year six days reached 50 or above, according to Accuweather records. Ten days hit 40 or above. In fact February's average high temperature was 36, 5 degrees warmer than the average.

Why is my brain troubled by this? Part of it is what the world actually looks like, and part of it is the unsettling imagination of what it might look like in the, maybe, not too distant future.

Snowfall totals for February were not abnormally low. A National Weather Service map showed that most of Maine north and east of Bangor got from five to ten inches, to as much as twenty inches of snow more than in the average

February. Waldo County and central Maine got roughly the average, or just less.

The difference this February appears to be that the snow melted into air after it fell. So the driveway gravel spent a lot of days weirdly exposed to daylight. Not to mention the brook, which in the natural history of Troy is almost always obliterated by plate ice for months, but this year was seen boiling along most of the winter.

Snowless February woods are sort of like a scene in a vaguely disturbing dream. You keep going about your business, but the atmosphere of your mind keeps wondering what's wrong with this picture, and what strange thing might happen next. You kind of feel like it could be anything.

And maybe it is: 2015 was the warmest year on record for Earth.[*] Before that it was 2014. The National Oceanic and Atmospheric Administration reports that the five warmest years out of the last hundred and thirty-five (i.e., since records have been kept) all occurred since 2005.

In December 2015, researchers studying satellite and other data reported that overall between 1985 and 2009, the surface temperature of Earth's lakes warmed "significantly, with a mean trend of 0.34° Celsius per decade." The report in *Geophysical Research Letters* — "Rapid and highly variable warming of lake surface waters around the globe," authored by dozens of climate scientists — states moreover that: "In both the Laurentian Great Lakes region [of North America] and in Northern Europe, lakes were warming significantly faster than the global average."

---

[*] Information in this paragraph is to the date of first publication of this article in 2016. At the time this book was in production in 2021, 2016 and 2020 were tied for warmest years ever recorded. The six warmest years worldwide in the 1880-2020 record all occurred since 2015; nine of the ten warmest years have occurred since 2005.

Whether Maine's lakes were part of the warming is not specified in the report. But I found a raw data sheet of lake water temperatures made available by the Maine Department of Environmental Protection to see if I could get a reliable clue. I couldn't, of course, because I'm not a climate scientist.

But my perusal, anyway, of surface temperatures recorded in mid-July and late August on Cobbosseecontee Lake and Unity Pond showed in both cases fluctuations up and down, year by year with no huge departures between the early 1980s and 2014. But in the 2000s the temperatures for both seemed to be generally a bit higher—more toward the mid 20s Celsius—than the temperatures in the 1980s, which seemed to land more toward the lower 20s. No July or August temperatures during the 2000s dipped to the teens, whereas one August day in 1982 Cobbosseecontee's temperature at the surface was 17.2 degrees, and one August day in 1986 Unity Pond's was 19.

Now, temperature data are not the stuff bad dreams are made on. Unless you start turning them over in your brain, which the scientists who made the global lake study did. After analyzing massive amounts of data, they concluded:

"The global average lake summer surface water warming rate found here implies a 20% increase in algal blooms and a 5% increase in toxic blooms over the next century ... as well as a 4% increase in methane emissions from lakes during the next decade. Increased evaporation associated with warming can lead to declines in lake water level, with implications for water security ..., substantial economic consequences ..., and in some cases, complete ecosystem loss. ... The widespread warming reported here suggests that large changes in Earth's freshwater resources and their processes are not only imminent but already under way."

March

To borrow an Irish novelist's trope about history, this winter has been a weird dream from which, try as we might, we may not awake.

Maybe I'll walk a turn or two around the snow-clear woods, and try to still my troubled mind.

A Crater in Troy

This happens pretty much every year. By the beginning of March there is so much snow plowed up around the yard we feel like we're living in a crater. From there, it only gets weirder.

To enter the driveway from Route 9, you pass between two large, pointed piles of snow. The ice-covered gravel track, such as it is, descends at what seems to be the same angle of slope I remember when I started into the Pyramid of Khafre in Cairo years ago and claustrophobia forced me to turn back. The track bends down over Bog Brook like a sort of miniature causeway, lined in winter by snowbanks, and bottoms out under snow-laden pine, maple, oak, cedar, ash, and other branches that close out low-angled winter sunlight and leave the ice there two to three miles thick. Or so we estimate. The driveway becomes a sort of tunnel. It then re-ascends for about fifty feet and comes out in the yard, which this past weekend was semicircled by seven-foot mountains of plowed-up snow. Welcome to Crater Troy.

If you look at these steep walls too long you start to feel uneasy. They're covered with rocky-looking frozen debris and peaked with boulders you imagine must be the same on the slopes of K2. The walls are all moon-ice white, closing off the world.

At one point I was standing in the middle of this driveway nowhere, looking up at the ice walls, and started thinking I'd seen this before. Rubble-strewn snowbanks or maybe glaciers, surrounding me. I wasn't sure if it was from a childhood memory, or a dream, or another lifetime. Trying to remember was like clambering along an underground passage. Then suddenly I was up the other side: the driveway crater is an exact similitude of Saturn's weird moon Hyperion.

Hyperion is a potato-shaped thing around two hundred twenty-four miles long and one hundred twenty-eight miles at its widest point. It's made mostly of ice. A spacecraft photo shows a huge crater gashing almost one whole side of it. The crater slopes are bright and steep and seem rubble-strewn. There are scores of other craters in the crater, with jagged-looking edges. It could be somebody's driveway in Troy.

How Hyperion got that way is essentially unknown. It could be the last big chunk of a larger moon that got blown apart in a collision or in a bombardment that Saturn's environs are thought to have undergone eons ago. But Hyperion (named at the time of its discovery in 1848 after one of the Titans, about whom not a lot is revealed in ancient Greek texts) has other shadowy inexplicabilities, too. For one thing, instead of rotating smoothly and evenly like other moons and planets, it tumbles chaotically end over end and side over side. It orbits Saturn every twenty-one days near the much larger moon Titan, which makes three rounds for every four of Hyperion's. Every time Hyperion passes into Titan's gravity, it jostles, changes its spin axis, and flails.

Hyperion's density is so low its interior is thought to be a webwork of ice caverns. If somehow you could make your way through its constantly skewing thirteen-day spin and set foot on it, not only would you be standing at the bottom of winter looking up at ice walls, but you might also descend through openings in the craters to minus 300 degree caves. You might crawl down shadowy tunnels and find frozen debris and tinkling landslide motions you wouldn't think possible. If you stayed down there long enough in that titanic darkness you might eventually wonder if it was an actual place or just a piece of winter tumbling end over end for the rest of eternity, or at least until the next bombardment of snow and ice. After a long time of freezing you inhabit this feeling of craterous remoteness and claustrophobia you want to crawl out of, and can't.

Winter lasts so long here, and its walls are so steep, and cold, and your mind, it wanders.

Backroad Battlegrounds

Something there is that doesn't love a road. In Maine, especially in March, when you get the uneasy feeling that winter will never loosen its jaws.

So many heaves and blast holes appear in paved surfaces at this time of year that the towns stock up on orange BUMP signs ahead of time, as if they were amulets that could ward off pothole elves. A stretch of road that was frozen flat and passable in the morning by afternoon appears to have been excavated. Ball joints crack and shock absorbers jounce. But it's not elves, exactly, it's that the road is caught in the tug of war between winter and summer. It always seems like winter's going to win.

It does not like to surrender. Its secret agents are cold and water (which may amount to the same thing), and it uses them in two ways: to create heaves and to create holes.

Frost heaves occur where ice lenses have formed underneath the road. An ice lens is a lens-shaped formation of ice. As winter digs in during December and January, soil and pavement freeze deeper and deeper. Ice fills the cracks and pores. After a certain depth, the soil remains warm enough not to freeze, and at the dividing line between the frozen and unfrozen soil, liquid water below is attracted upward through nooks, crannies, and capillaries in the soil to ice pockets above.

How this occurs, exactly, is not conclusively understood. Strange things are happening in the water down there. Some of it is actually "supercooled," meaning it's colder than the freezing point, yet still liquid. The supercooled water freezes on contact with the ice pocket, adding to it. It grows in the shape of a lens and sends the frozen ground swell under the pavement, exerting pressure upward, and then heaves, humps, dips, and gaps form in the road that sometimes two vehicles can pass abreast. Slowly.

While frost heaves are created from the bottom up, pot-holes are created from the top down. When ice melts on warmer days it leaves gaps in the pavement, and liquid water running off the snow and ice seeps in. At night winter seizes its opportunity and drives the temperature down again, icing everything. Freezing water expands naturally by 9 percent in volume, so there's more ice pushing inside the pavement. When the March sun rises, higher and warmer than it was in February, more ice melts and more moisture gets into the gaps. The pavement expands, contracts, gets stressed by the ice, and cracks. As cars and trucks rumble over, the pavement breaks up and tires kick out pieces and form holes. Heavy trucks are winter's special allies in demolishing pavement. So there are at least two things that don't love a road.

The war between winter and summer goes on for what seems like a lifetime every year. Despite the fact the sun is getting higher in the sky, in March the prospects for victory seem bleak, especially on places like the Rogers Road in Troy, one among many that could have been an able competitor in the Worst Road in Maine contest. (Which is a real thing.)

Some think the world will end in fire, but right now ice seems great and will suffice. Suffice to bottom out any vehicle that is not a skidder, anyway.

The Auroras of Spring

In March, as the sun shows distinct signs of perking up due to the Earth swinging the Northern Hemisphere back into more direct sun rays, another annual occurrence takes place that fascinates and partly stumps the astronomers. Around the vernal equinox, about March 21, auroras start firing more intensely in the northern sky.

The less the astronomers understand something, the more it fascinates them, it seems. And in this I feel a connection to them. I've stood on the edge of a field in Troy in past years and watched auroras fill the sky—huge snakes of polar green kindling like ice and fire between me and the Big Dipper. They twist and seem to shimmer with blue and violet tinges, then fade away like thoughts. And then re-emerge. The farther north you go, the brighter and more enormous they are. You can only think how the imagination of an ancient Abenaki native standing in a clearing might have been struck by the huge cloud mountains running like water through waves of light. What did they signify?

Nowadays we explain the auroras like this: the sun generates streams of electrically charged atomic particles called the solar wind. Some of the particles moving us-ward get trapped in the Earth's magnetic field, and they tend to flow toward the poles. There, they interact with other particles, energy is released, and arcs of green and rose light, stupendous to our eyes, slowly bend and twist across the sky.

Around the vernal equinox, when the Northern Hemisphere has reached the halfway point between its full tilt away from the sun in December and its full tilt toward the sun in June, the auroras intensify. The scientists do not know why, exactly. It has nothing to do with temperatures. It's thought that "rope-like magnetic connections" develop along the angle between the Earth and sun, and more charged particles travel along those connections and channel toward the magnetic pole. In turn, the northern lights intensify.

On the sun, flares, prominences, and ejections of material which amount to colossal explosions take place pretty frequently. Some of them are associated with sunspots, and sometimes the explosions are so huge they disrupt radios and electric grids on Earth. Their causes are not understood well enough to predict when they're going to happen in most cases. Sunspots gather and desist in eleven-year cycles, though, and so auroras tend to be more abundant and show brighter "auroral substorms" around the solar maximum.[*]

The sun and Earth have invisible connections; this much is known. Some of those connections take visible shape from time to time as tremendous green serpents of light in the sky. And those connections intensify—for reasons not fully understood—just when spring and winter are haggling over March, which is the tipping point toward summer. They're a sign summer will indeed blaze up, as quick as a straw fire, or as an auroral substorm, or as a thought. Afterward, too, come the auroras of autumn. What significance this has, exactly, I have no idea. Except to say their frigid brilliances, their blue-red sweeps, their polar green, are awesome, which is a clue of some unknown kind to some unknown thing.

---

[*] A solar minimum of sunspot activity was just concluding around the time this book went to press in 2021. The next solar maximum is around 2025.

The Dog of Winter

It's March.

The daytime temperature has been in the vicinity of 35 to 45 degrees with biting wind. A foot of snow fell out of the sky a couple of weeks ago. The ice mountains at the top of the driveway, which around the end of February had started showing hopeful signs of receding as they do every year, built back up in geological fast-forward so the backyard once again resembles central Antarctica.

Winter is like a mean dog. It gets its teeth in your leg and won't let go. Starting in late February when the sun is suddenly higher in the sky and stays there longer, the dog nods off a little during the day. But you have to keep a healthy mental distance because at night it wakes up and growls. March nights refreeze the mud that plagued the driveway in the afternoon. It always snows. In March 2005, more than thirty-two inches of snow fell on Portland. In 2001, 40.5 inches. In 1956, more than forty-six inches. In 1993, forty-nine. Those icy fangs.

One year the dog fell asleep in mid-March and the temperature was in the 90s for a few days. You could see it coming: people thought—or hoped so hard they believed—that winter had left early. But a week later, as I recall, a pervasive sense of dejection set in when the cold nights re-awoke and cranky winds snarled and snapped again at our ankles.

March is the cruelest month in this part of the world, but April can be its own cur. On April 4, 1954, it was 8 degrees in Portland. On April 27, 1964, there was a 49-degree swing in temperature. The coldest April averaged 38.4 degrees in 1943.

Even in 2012, the second warmest April on record up to then, the mean temperature for the month was only 47. In the first week of April 1982, 15.9 inches of snow fell on Portland. And that's in Portland, which to us up here in the woods is off the subarctic map and into a temperate zone. In April 2011, 15.4 inches of snow fell on Bangor. When I taught at Unity College back in the last century, I remember Earth Day celebrations being driven indoors by wet, raw bags of snow falling out of cold gray skies.

Here in Troy, as my ultra-reliable plow guy Jason (who lives just across the pond in Unity Plantation) observed, "You guys seem to get more snow than everybody else." In the big storm in February 2013, when towns were reporting snowfall in the twenties of inches—which is enough already—I measured thirty inches of new snow straight up on my deck. In the insomniac dog winter of 2011, clawlike plates of ice in the driveway seemed to be permanent topological features by late April and shreds of our Antarctic mountains held on into the first week of May.*

"Though the frost is nearly out of the ground," Thoreau wrote in his journal on March 30, 1852, "the winter has not broken up in me. It is a backward season with me. Perhaps we grow older and older till we no longer sympathize with the revolution of the seasons, and our winters never break up."

My way of saying the same thing, twenty-seven Marches further on than Thoreau was when he wrote this at age 33, is that winter's teeth do not let go willingly, hereabouts. As someone told me years ago, God invented March in Maine so people who don't drink know what a hangover feels like. Or a mean dog, who just won't go to sleep.

---

* This turned out to be less and less unusual—or alternatively expressed, usual—in following Aprils.

My Old Friend Leo

One night I was making my way across the Arctic wasteland of my driveway and was startled to look up and see my old friend Leo.

Startled because this usually doesn't happen until about mid-March—and going by the empirical facts (snowbanks six feet high and 6 degrees Fahrenheit out my kitchen window that morning), it was still February. But wait. The first day of spring had already come and gone?

Somehow winter has been in slow motion in recent years, and stretching out of March into April. But no matter what the cold and snow do to your mind, the sky keeps strictly to the schedule. Which probably explains why part of me was so glad to see Leo. It really is getting on toward spring.

Leo, the Lion, has stalked the evening sky for months, actually, but by around 10 p.m. in the middle of March it dominates my southern treetops.

On a clear night, look south and you can pretty easily spot six bright stars roughly in the shape of an anvil. To the left at the narrow end is the bright star Denebola, our foreshortened version of the Arabic phrase *dhanab al-asad*, the lion's tail. At the lower right of the anvil is Regulus, the brightest star in Leo. In the imagination's star-lion, Regulus is at the joint of the front legs. Directly above Regulus is Eta Leonis (with no common name of its own, oddly, just the Greek letter η), in the lion's chest, and above and slightly left is Algieba (*al jabbah*, the mane), at the shoulder.

If you track your eye above Algieba, you'll see three more stars, not quite so bright, looping in a sort of backward question mark; these outline the lion's head. Along the lion's back, the star between Algieba and Denebola is Zosma, which sounds like a name out of some rambling fantasy trilogy and is actually a misapplied version of a Greek word for girdle. Arab astronomers more appropriately called it *Al Thahr al Asad*, the lion's back.

If you imagine the sun traveling a great circle every year across the sky, that road—called the ecliptic—runs through the twelve constellations of the Zodiac (plus one other, Ophiuchus). Leo is one of the twelve. To its east lies Virgo (still mostly down behind my trees on March evenings). To the west is Cancer, the Crab, made of four dimmer stars.

Like all the constellations, Leo also has a lot of interesting objects that you can't see with your naked eye. But with a little help from the astronomers and their large telescopes and calculations, your mind's eye can picture them.

Some are galaxies. About a third of the way from Regulus to Denebola, and down a bit, is the Leo I group of galaxies. They're part of the Virgo Supercluster and are in the vicinity of eight hundred and twenty thousand light-years from us, too faint to see without a telescope, of course. A little farther toward Denebola is the Leo Triplet group of galaxies, or M66 group, around thirty-five million light-years away.

Directly south of the Leo I group is the star Wolf 359, which also is invisible to us because it's a red dwarf shining at a very dim magnitude 13.5. But it's interesting to play with in thought because it's the fifth-nearest star known, about 7.8 light-years away.

At the other extreme, Eta (η) Leonis, bright in the anvil, is a supergiant, far, far away but so big we see it easily. The estimates vary, but this star is somewhere around twenty-five times as large as the sun and has a luminosity of five thousand six hundred suns, or some think up to nine thousand five hundred suns. Robert Burnham, whose *Celestial Handbook* has a more or less scriptural presence among starwatchers, estimates it could be as luminous as thirteen thousand suns. Eta is somewhere between thirteen hundred and eighteen hundred light-years away from us; if it were 32.6 light-years from us (a standard used to compare stars' brightnesses, called the absolute magnitude), it would outshine even Venus in the sky.

Southeast of Regulus is another supergiant you can see, Rho (ρ) Leonis. Rho is estimated to be from three thousand six hundred and fifty to five thousand light-years away and has a luminosity of one hundred sixty-five thousand or maybe up to two hundred ninety-five thousand suns. To us it shines at magnitude 3.85, well within our naked-eye sight, but its absolute magnitude is probably around minus 5.7, even brighter than Eta.

One more for the road of your imagination: to the west of Regulus and slightly up is the star R Leonis, which sometimes you can pick out with your eye and other times you can't. It's not just your imagination. R Leonis is a variable star, meaning its brightness changes over time. In R Leonis's case, the brightness changes radically over a long period of three hundred and twelve days, from magnitude 5—which you can see on a dark night if you know where to look—to magnitude 10.5, which is fainter than some of the galaxies in the Leo I and Triplet groups.

When I say Leo is an old friend, I'm speaking metaphorically, I guess. Every year, just as spring is breaking out of winter this old friend appears like spring out of the driveway Arctic. Right there by the house.

Why Is It So Cold in Winter?

Our harsh Northeast winters usually keep beating up on us into April. On April Fools' Day 2011 I measured an even sixteen inches of snow that had fallen on my deck the night before.

Why is it so cold here?

Well first of all, the seasons come and go because the Earth is tilted in relation to the sun. As we circle around every three hundred sixty-five or so days, the tilt alternately takes the Northern Hemisphere into and out of the sun's directmost rays. In other words, during summer the sun's rays strike us more or less directly; but in winter they're at an angle, and for fewer hours, heating us up less. The farther north you go, the less direct sun you get and, in general, the colder it is. In the summer, we tilt more toward the sun, catch more direct rays for longer each day, and are warmer.

That accounts for summer and winter, but it doesn't account for some perplexing variations in temperatures across the latitudes of the Northern Hemisphere. When I visited Iceland in a fit of youthful insanity a few decades ago, I was shocked to learn that the average wintertime temperature in Reykjavik is higher than New York City's, even though Reykjavik is just above latitude 64 (the Arctic Circle itself is just a hop, skip, and jump away at latitude 66 degrees, 33 minutes), while New York is just above latitude 40. And yet, the average low temperature in January for Reykjavik is 27 Fahrenheit, while in New York it's 26. In Bangor, which is at latitude 44 degrees, 48 minutes, the average low January temperature is 8.

How could Reykjavik be almost 20 degrees of latitude farther north and 20 degrees Fahrenheit warmer on January nights than Bangor?

Well, that's not all. Most of Europe is farther north than us, and warmer in winter. At almost exactly the same latitude as Bangor is Bordeaux, in southern France, where the average January low temperature is 36 and the average high 49. Bangor's average January high is 26. London, England, which is just above latitude 51 degrees north, has an average January high of 46 and low of 36, while Labrador City, Labrador—which must be one of the world capitals of cold if you ask me—at latitude 52 degrees, 56 minutes has an average January high of 2 and low of minus 18. Hexham, near the ancient Roman wall in northern England where I've spent some time, is farther north than Labrador at latitude 54, but its average January high temperature is 38. That's *plus* 38. In Labrador, farther south, it's 2.

What is going on here? The Gulf Stream carries warm waters from the balmy Gulf of Mexico up past Newfoundland and washes around Iceland, Britain, and the shores of France and Portugal, moderating air temperatures there year round. But that turns out to be a relatively minor factor. A greater impact is sort of the inverse.

The warmer water flowing along the East Coast toward Europe doesn't heat Britain, France, and Iceland as much as it freezes us. It warms the air above it and sets up atmospheric waves that actually suck cold air down from Arctic Canada to descend on the Northeast, encoldening us beyond our neighbors to the east. Interestingly, a similar process works in the western Pacific, accounting for the similarly harsh winters experienced in China's northeast.

So that goes a long way to explaining it, but as usual the explanation doesn't ease the beating. I've been in London during February and seen daffodils cheerfully abloom, to no one's surprise there. Yet in April and 7 degrees of latitude farther south in Troy, my old blue Mercedes has sometimes remained buried in snow. The average high temperature for Bangor in April is supposedly 53, and the low 33. In the several Aprils I've spent in London, I remember no snow. Here, though, you've got to keep your mind right. Winter ain't over till it's over.

Before

Get Your Mind Right

While everyone else is eagerly forgetting winter, our yard in the latter part of April is still ruled by Big Mud, which leaks moisture through the basement walls and impedes even the thought of walking up to get the newspaper. Often in April we are still imprisoned by three, four, five foot snowbanks surrounding the driveway. For all intents and purposes, it's March here for most of April.

There is usually, of course, a run of springlike days. But it still goes to freezing or below at night, the cars still need to be washed, and we often see no sign of Persephone before May, when the last of the inner-woods snow finally disappears. And there's always some bleak wind that makes you think twice about which part of the seasonal cycle you are actually inhabiting. The full-on sense that summer has arrived doesn't usually hit until June. About six weeks later in August there start to be some chilly—if beautiful—evenings and a few orange leaves that indicate the great packing up is already underway. Summer is as short as you think.

By the calendar, winter should start showing signs of opening the cell door around the end of March. Vernal equinox spring arrives around March 21 every year because the angle and path of the Earth with respect to the sun take us into it with planetary precision. But March's average temperature in 2011, for example, was 26 degrees, 4 degrees below normal average. The March before was even colder, 24 degrees. The average through mid-April was 39 degrees, 4 degrees below normal. The first two weeks of April averaged just 37.

But spring—that "memorable crisis which all things proclaim," Thoreau calls it—in practice is less an astronomical date and more a state of mind. Toward the end of the annual confinement in winter's box, you orient yourself to the snow depth, the height of the sun at midday, and the outside thermometer by measuring in your mind when you'll be able to take an ice-free walk.

I always expect Big Mud to ransack the driveway in about mid-March. There can be some backbiting April weather (cf. the blizzard of 1982—during which I was luckily in Greece) and wet heavy snow often seems to fall on Earth Day, but in early Aprils past the driveway would be drained and by mid-month bright yellow forsythia blossoms would be flourishing. My inner clock up to now has been caught in a cycle of looking for it all to happen on this timeline.

But for some years now, the vernal crises have happened later than this. Not only has winter been locking us down well into April, but it's been starting later too. Before the snowpocalypse descended, the earth was still bare in January, as if December had delayed a month. By April, we have usually not yet cleared March. The whole winter has jogged on almost a month.

This gets exasperating, until you realize the problem is less a jogging climate, and more your own disposition to it. What we have here is failure to communicate with nature. Projecting your idea of spring onto March is what causes you the friction.

To escape the exasperation cycle, you have to get your mind right. Next winter, I'm betting, March will still be winter, like the last few years. Instead of reliving the same bad karma when Big Mud is yet again delayed to April, I'm calmly accepting that March is February and April is March. May will probably cram April and itself into thirty-one days, the black flies will mount their annual raid, and then June will blossom.

Next winter, I'll bet, Persephone will retreat back into her cottage later than you expect, and sashay back out a little later past the equinox than you wish she would. I'm expecting nothing more.

And as King Lear presciently, eerily observed, nothing will come of nothing.

About the Author

Dana Wilde lives in Troy, Maine, and writes the Backyard Naturalist column which appears regularly in the *Kennebec Journal* and *Morning Sentinel* newspapers, and originated in years past as the award-winning Amateur Naturalist column in the *Bangor Daily News*. He has been a college professor, editor, Fulbright scholar, and NEH fellow. He holds a bachelor's degree from the University of Southern Maine and doctoral and master's degrees from Binghamton University, where his doctoral dissertation, *Ursa Major: Essays in Outer Space*, was recognized as the Distinguished Dissertation in the Humanities and Fine Arts for 1995, in part for its efforts to bridge the "two cultures" of science and the humanities. His writings appear regularly in *The Working Waterfront* newspaper, and have appeared in many popular, literary, and scholarly publications, including *The Magazine of Fantasy & Science Fiction*, *The Quest*, *Alexandria: Journal of the Western Cosmological Traditions*, *Mystics Quarterly*, *Exquisite Corpse*, *North American Review*, *The Maine Entomologist*, and many others.

Books

A Backyard Book of Spiders in Maine (2020)
Summer to Fall: Notes and Numina from the Maine Woods (2016)
Nebulae: A Backyard Cosmography (2012)
The Other End of the Driveway (2011)
The Big Picture (1983)